SUMMER ACTIVITIES for Fall Readiness

FIRST GRADE

Jeanine Manfro

GARLIC PRESS
Educational Materials for Teachers and Parents

Printed in USA

GARLIC PRESS
Educational Materials for Teachers and Parents
899 South College Mall Road
Bloomington, IN 47401

www.garlicpress.com

Publisher: Douglas M. Rife
Author: Jeanine Manfro
Art Director: Michelle Ste. Marie
Illustrators: Joanne Caroselli, Sherry Neidigh, Emilie Kong, Michelle Ste. Marie
Interior Design: Joanne Caroselli
Cover Design: Joanne Caroselli

ISBN 978-1-9308-2055-5
GP-202
TPS, Newton, IL 05 13

Table of Contents

Introduction

Summer is a time when families can relax and enjoy lots of great adventures together. It is also a time when many children forget much of what they learned during the previous school year. This workbook was developed to help children maintain the academic skills they have already acquired and to get a head start on the skills they will learn in the coming school year.

COMMON CORE STATE STANDARDS

Summer Activities for Fall Readiness supports the Common Core State Standards Initiative (CCSS). These standards were developed by teachers, administrators, and educational experts nationwide to ensure that all students are provided with a top-quality education. The standards have been adopted by nearly every state and they provide clear and consistent information about what students should learn at each grade level.

SHORING UP SKILLS AND SETTING SAIL

Help your child sail into first grade with confidence! With just 10 to 15 minutes a day, this workbook can be completed over the summer and your child will enter the school year ready to learn.

The book is divided into 10 one-week sections. Five reading/language arts pages and five math pages are provided for each week. The first half of the book is dedicated to reviewing Kindergarten skills and the second half provides a sneak-peek at what your child will learn in first grade. Pages are clearly marked so that you'll know at a glance if your child is working on a **Kindergarten skill** or a **First Grade skill**.

At the bottom of each page, there is a suggestion for an additional activity that you and your child can complete together to extend the learning even further.

Here are just some of the **Common Core State Standards** presented in this workbook:

Reading/Language Arts
- Writing Uppercase and Lowercase Letters
- Phonics
- Rhyming Words
- Sight Words
- Nouns, Verbs, and Adjectives
- Capitalization and Punctuation
- Reading Comprehension
- Understanding Informational Texts

Math
- Writing Numbers
- Counting to 120
- Addition and Subtraction
- Word Problems
- Comparing Numbers
- Time and Money
- Place Value
- Using Data

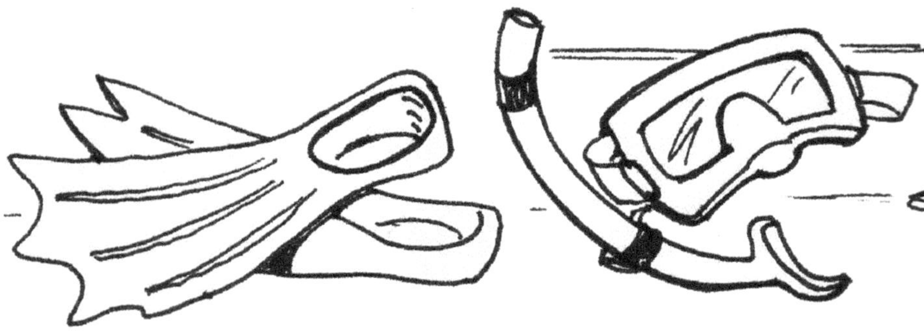

A Sea of Stories

Reading aloud to your child is perhaps the best way to prepare for first grade. Here are some wonderful books to get you started!

FABULOUS FICTION

Aesop's Fables by Jerry Pinkney. Chronicle Books, 2000.

Animal House by Candace Ryan. Walker Books for Young Readers, 2010.

Crustacean Vacation by Brian Benoit. Islandport Press, 2012.

The Duckling Gets a Cookie!? by Mo Willems. Hyperion, 2012.

Fairytale News by Colin and Jacqui Hawkins. Walker Children's Books, 2005.

How I Became a Pirate by Melinda Long. Harcourt, 2003.

Junie B. First Grader (At Last!) by Barbara Park. Random House, 2002.

Knuffle Bunny: A Cautionary Tale by Mo Willems. Hyperion, 2004.

Sally and the Some-Thing by George O'Connor. Roaring Brook Press, 2006.

This Plus That: Life's Little Equations by Amy Krouse Rosenthal. HarperCollins, 2011.

The Three Little Wolves and the Big Bad Pig by Eugene Trivizas. Margaret K. McElderry Books, 1997.

Z Is For Zombie by Merrily Kutner. Albert Whitman, 1999.

KNOCK-OUT NONFICTION

About Habitats: Grasslands by Cathryn Sill. Peachtree Publishers, 2011.

All the Water in the World by George Ella Lyon. Atheneum, 2011.

Caterpillars by Marilyn Singer. EarlyLight Books, 2010.

Cowboys and Cowgirls: Yippee Yay! by Gail Gibbons. Little Brown, 2003.

Even an Octopus Needs a Home by Irene Kelly. Holiday House, 2011.

How Did That Get in My Lunchbox? by Chris Butterworth. Candlewick, 2011.

Just a Second by Steve Jenkins. Houghton Mifflin, 2011.

A Koala's World by Caroline Arnold. Picture Window Books, 2008.

A Little Peace by Barbara Kerley. National Geographic Society, 2007.

Me on the Map by Joan Sweeney. Dragonfly Books, 1998.

The Quilting Bee by Gail Gibbons. HarperCollins, 2004.

What Do You Do with a Tail Like This? by Steve Jenkins. Houghton Mifflin, 2003.

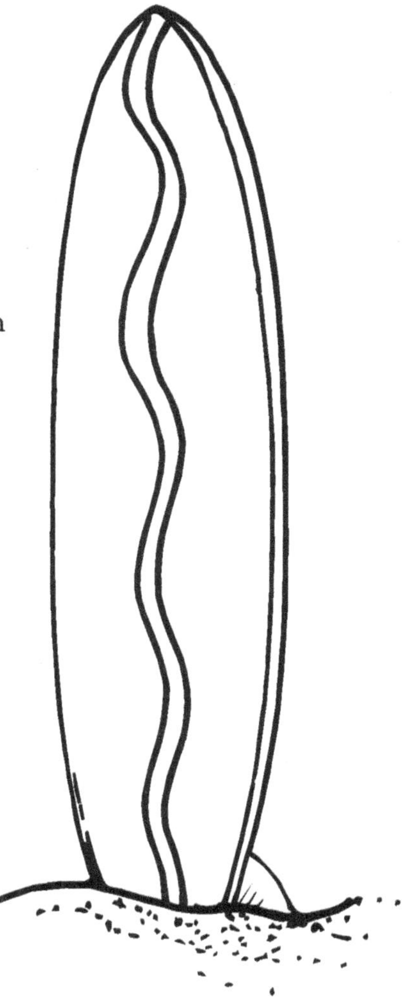

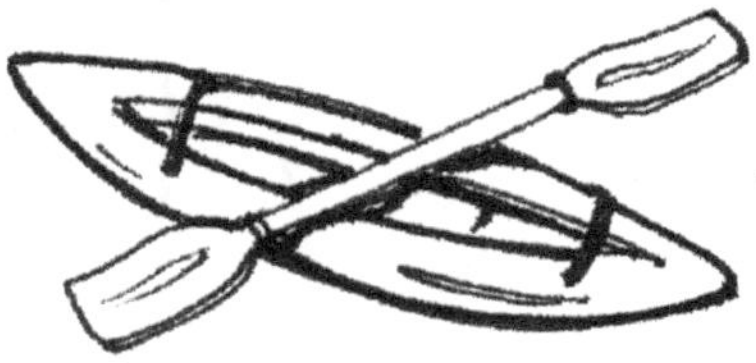

A Boatful of Books

MONTH

SUNDAY	MONDAY	TUESDAY	WEDNESDAY	THURSDAY	FRIDAY	SATURDAY

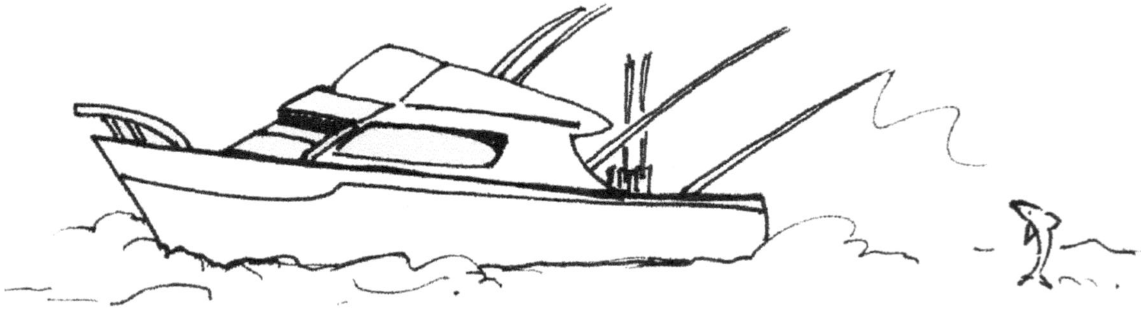

Make three copies of this page—one for June, one for July, and one for August. List the books that you read to your child during each month. Make it your goal to read together every day. Plan a special reward for the end of the summer when you reach your goal.

Fall Readiness Checklist

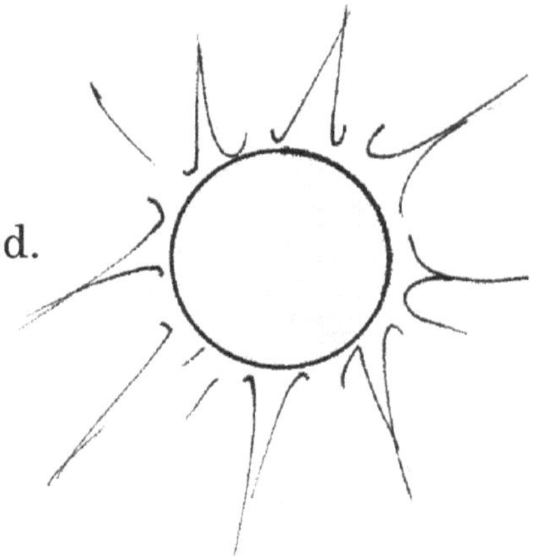

Children entering first grade should be able to do these things. Check the box for each skill your child has mastered.

READING /LANGUAGE ARTS SKILLS

- ☐ Know the names and sounds of individual letters and some letter combinations.
- ☐ Read common sight words.
- ☐ Use rhyming words.
- ☐ Capitalize the first word in a sentence.
- ☐ Use a period to end a sentence and a question mark to end a question.
- ☐ Ask and answer questions about details in a text.
- ☐ Identify characters, settings, and major events in stories.
- ☐ Identify the main topic and retell key details of an informational text.

MATH SKILLS

- ☐ Count to 100 by ones and by tens.
- ☐ Write numbers 0 to 20.
- ☐ Compare numbers and sets of objects using "greater than" and "less than."
- ☐ Add and subtract within 10.
- ☐ Solve addition and subtraction word problems.
- ☐ Understand the basics of place value, using tens and ones to describe numbers to 20.
- ☐ Name the days of the week.
- ☐ Tell time to the hour.
- ☐ Identify shapes.

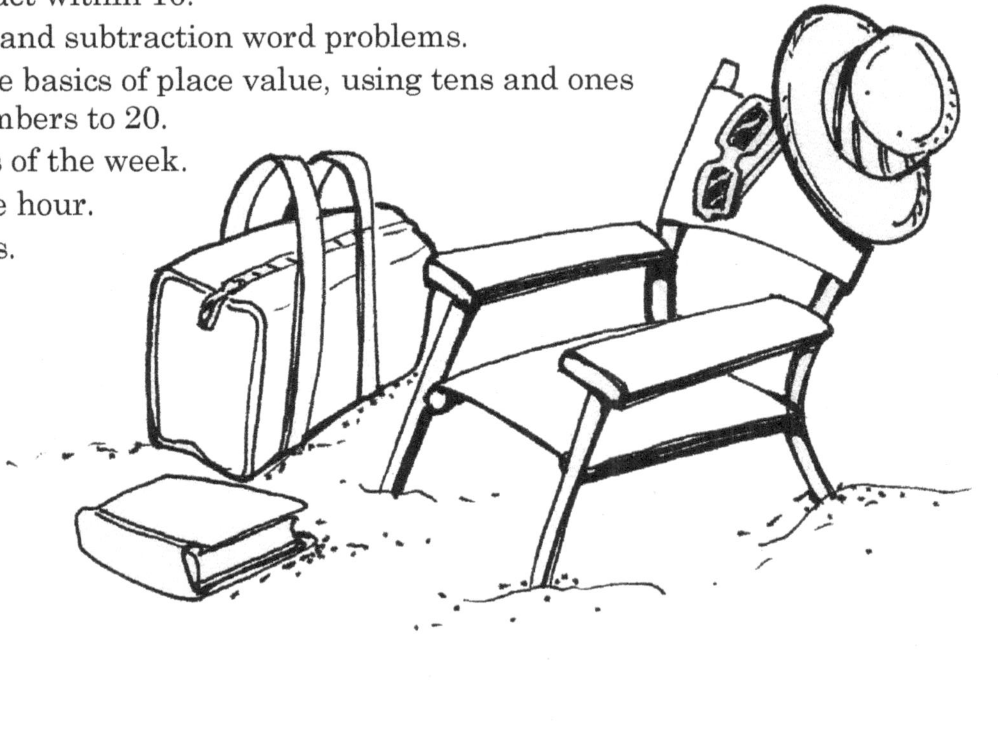

First Grade Words

Here is a list of words that first graders may learn to read and write. Some of these words may have been learned in Kindergarten. Others may be taught later in the first grade year. The words are based on the Dolch list of frequently used words. Have your child practice reading and writing some of the words from this list each day.

after	eat	know	put	this
again	every	let	ran	too
all	fly	like	ride	under
am	four	live	round	walk
an	from	may	saw	want
are	get	must	say	was
as	give	new	she	well
ask	going	no	so	went
at	good	now	some	were
ate	had	of	soon	what
be	has	old	stop	when
black	have	on	take	white
brown	he	once	thank	who
but	her	open	that	will
by	him	our	them	with
came	his	out	then	yes
could	how	over	there	
did	into	please	they	
do	just	pretty	think	

Lost

Read the two nursery rhymes.

Little Betty Blue,
Lost her holiday shoe.
What will poor Betty do?
Why, give her another,
To match the other,
And then she will walk in two.

Little Bo-Peep has lost her sheep,
And can't tell where to find them;
Leave them alone,
and they'll come home,
Wagging their tails behind them.

Talk about the rhymes.

1. What is the name of the character in each nursery rhyme?
2. How are the characters alike? How are they different?
3. What did each girl lose?
4. How do the girls solve their problems?
5. Both rhymes are told in two parts. Who do you think is telling the first part of the rhyme? Who do you think is telling the second part?

Have your child draw a picture of both characters and write a sentence about them.

Seashells

Count the seashells. Write the number five times each.

1. 

2. 

3.

4. 

5. 

6. 

Use a staircase to have your child practice counting. As your child moves up the stairs, have him or her count up. Then have your child practice counting backwards as he or she moves down the stairs.

ABC's

Trace and write the uppercase letters.

A B C

D E F

G H I

J K L

M N O

P Q R

S T U

V W X

Y Z

Have your child "rainbow write" the uppercase letters. After writing each letter once, have your child trace it again and again with different colored crayons.

Number Nests

Count the eggs. Write the number five times each.

1. 

2. 

3. 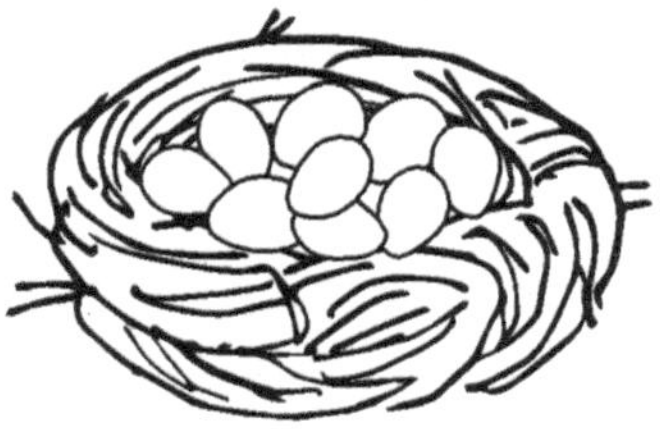

4. 

5. 

Number the cups in an egg carton from 1 to 12. Have your child count out quantities of small items such as marbles, o-shaped cereal pieces, or pennies, and place the sets in the egg cups with the corresponding numbers.

Read the story. Then answer the questions.

Our Flag

This is our flag. It has 13 stripes. The stripes are red and white. There is a blue rectangle in the corner of the flag. There are 50 white stars in the rectangle. The stars are for the 50 states.

1. How many stripes are on the flag? ____________________

2. Why are there 50 stars?

____

3. What three colors are on the flag?

____

Draw a picture to show one place where the flag is flown.

Teach your child some of the etiquette rules for the flag. For example, the flag should not ever touch the ground and if it is flown at night, it should have a light shining on it. Explain that following these rules demonstrate respect for the flag and the country.

Lovely Ladies

Count the spots on each ladybug. Write the number five times.

1. 

2. 

3. 

4. 

5. 

Cut sheets of poster board into 16 long strips. Number the strips from 0 to 15. Use a dot marker to put a matching number of dots in the middle of each strip. Finally, write the number word for each strip in the last section. Cut the strips into thirds and mix them up. Challenge your child to find the three matching pieces for each number.

Little Letters

Trace and write the lowercase letters.

a b c

d e f

g h i

j k l

m n o

p q r

s t u

v w x

y z

Write an “a” on the sidewalk with chalk. Then have your child write “b.” Continue to alternate until writing letters until you reach the end of the alphabet.

Dandy Dominoes

Count the dots on the domino tiles. Write the number five times.

1. 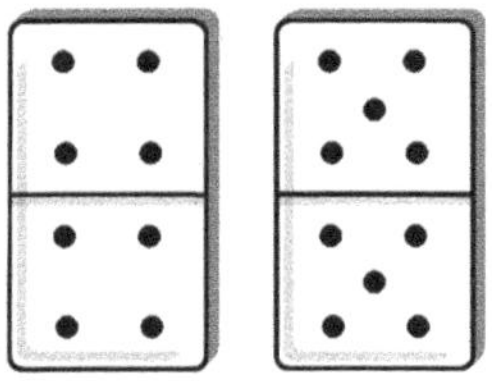

2. 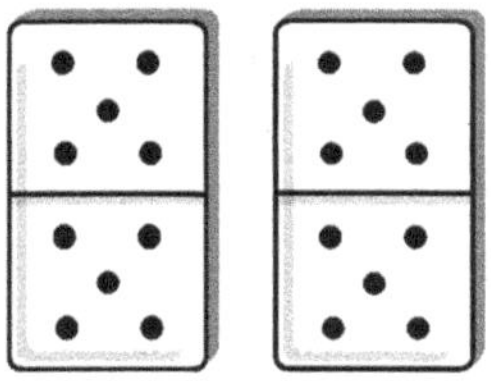

3. 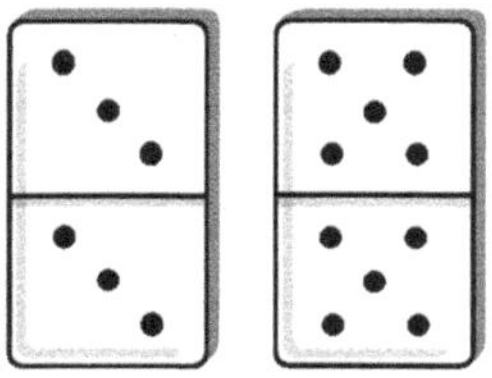

4. 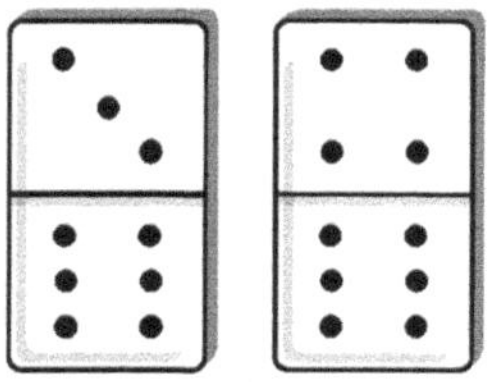

5. 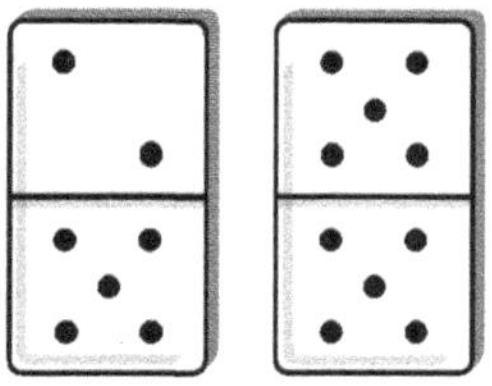

Have your child roll two dice and add the numbers shown. Then have him or her practice counting from that number up to 20. On the next roll, have your child count backwards from the number rolled down to zero.

Read the poem. Then answer the questions.

Little Band

Rum, pum. Rum, pum.
Carly beats on the drum.
Toot, toot, toot, toot.
Colin blows on the flute.
Little band wants to play
And have us dance through the day.

1. What sound does the drum in the poem make?

2. Who plays the flute?

3. Write a sentence that tells what the band wants.

Have your child use recycled materials to make a unique instrument. Invite him or her to play the instrument for you.

Engine, Engine, Number Nine

Fill in the numbers on the train cars.

1.

2.

3.

4.

Play this fun outdoor counting game with your child. Designate a starting line and a finish line. Then tell your child to estimate how many steps it will take to get to the finish line. Have your child count the steps to check the estimate. Repeat with various types of steps. Try giant steps, baby steps, hops, etc.

Read the story. Then answer the questions.

Picnic Time

For the picnic, bring the 

And the 

, if you please.

Bring the 

and some 

.

Bring the 

to fill us up.

1. What event is being planned in this poem?

2. What will be the drink?

3. If you went on a picnic, what foods would you bring?

Have a picnic with your child. Let him or her help prepare the food.

Count With Me

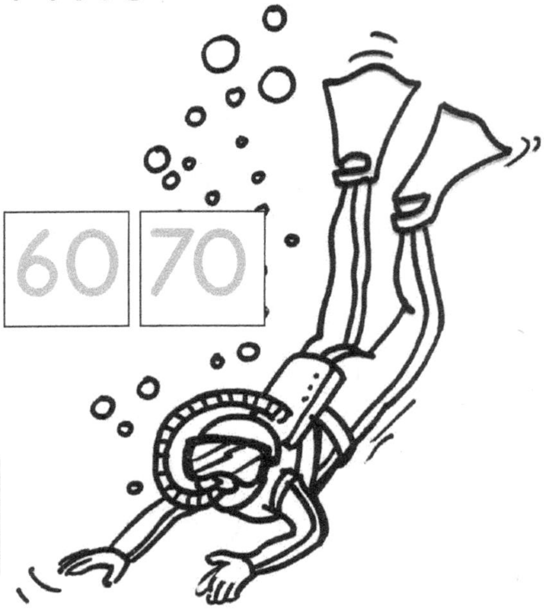

Count by 10.

1. Count by 10 to 70.

10	20	30	40	50	60	70

2. Count by 10 to 50.

			40	

3. Count by 10 to 90.

	20							

4. Count by 10 to 100.

		30							

5. Count by 10 to 40.

10			

6. Count by 10 to 80.

				50			

Have your child take off his or shoes and head outside with a pan of water. Tell your child to step into the water with both feet and then jump across the sidewalk to make pairs of footprints. Challenge your child to count by tens to find out how many toes are shown on the sidewalk before the footprints dry.

Rhyme Climb

Read the word on each ladder.
Write three words that rhyme with it.

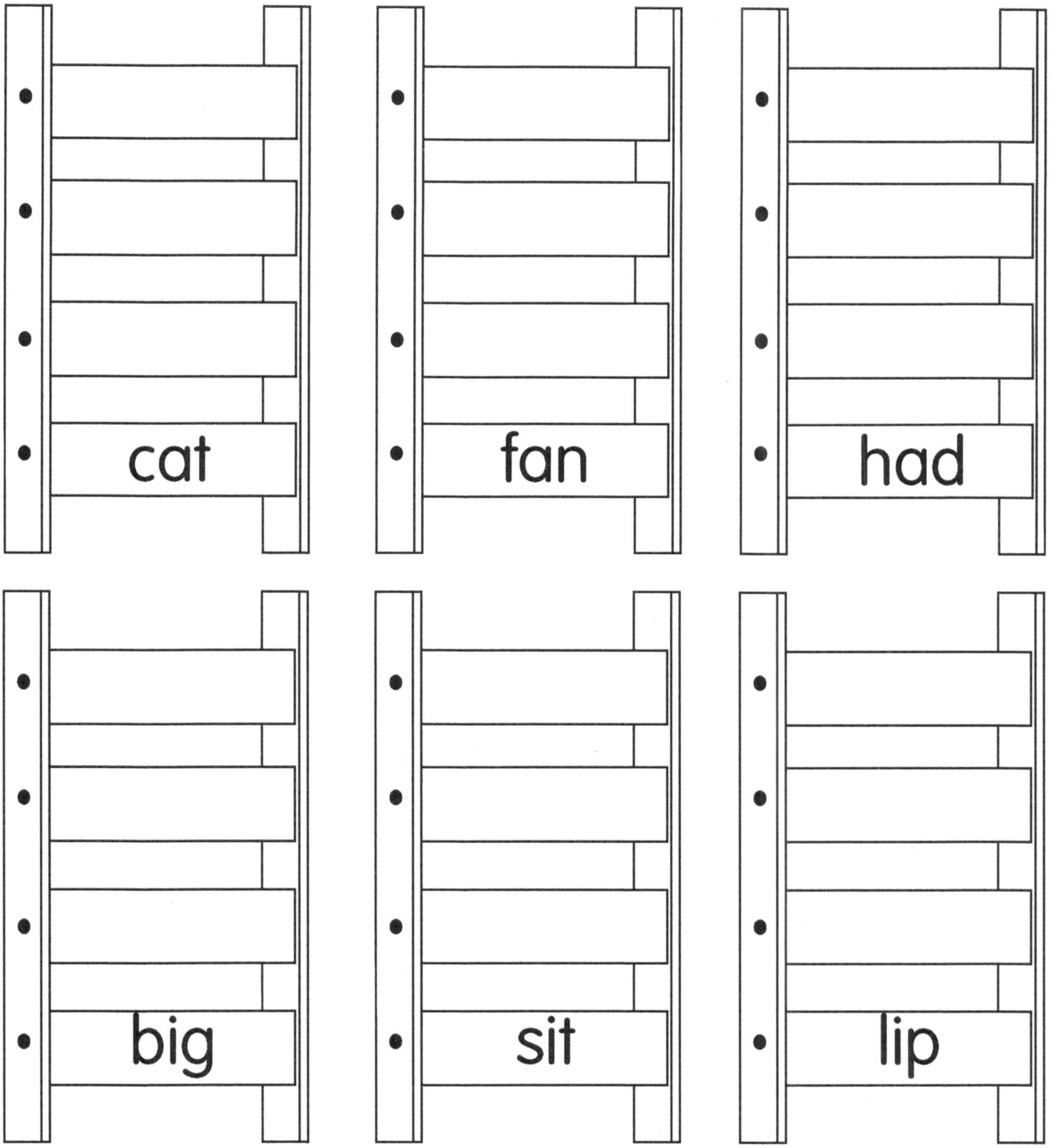

Read the Dr. Seuss book, *There's a Wocket in My Pocket*, with your child. Then encourage your child to come up with silly nonsense words that rhyme with things found in the house, such as *bindow/lindow/window, jair/zair/chair, etc.*

High Five

Count by fives to tell how many fingers in all.

1. 

5 10 15 20 25

2. 

3. 

4. 

5. 

Give your child some cereal loops that come in different colors. Have him or her string the loops on a length of yarn, changing colors after every fifth loop. When your child has 100 loops on the yarn, tie the ends together. Then have your child practice counting the loops by fives.

Read the passage. Then answer the questions.

Plants

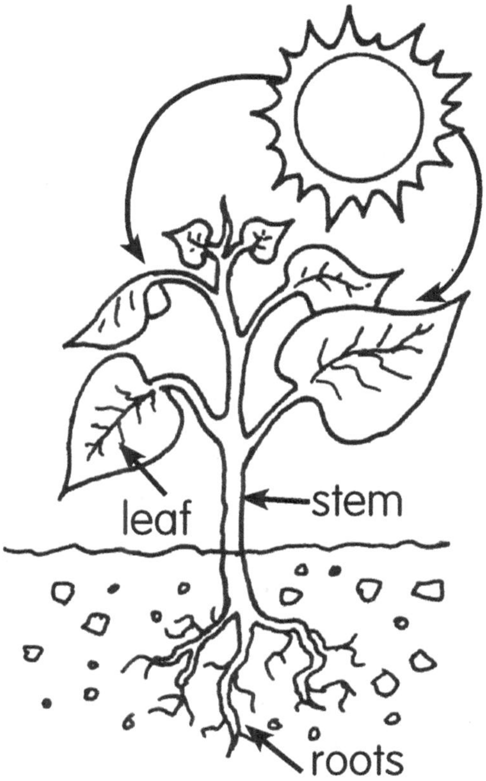

Plants are living things that make their own food. To do this, they use their leaves to soak in the sun's energy. Plants also use their roots to take in water. Many plants need sun, water, air, and soil to grow and make new plants.

1. What can plants make?

2. What do plants use to get the sun's energy?

3. How do plants take in water?

4. Which of these things does a plant not need?

water sun clothes soil air

Place a celery stick in water dyed with food coloring. Ask your child to predict what will happen to the celery. Have him or her draw pictures of what is observed over several days.

Lucky Duckies

How many ducks are there? Count by twos.

1.

2 4 6 8

2.

3.

4.

Let your child use a rubber stamp and an ink pad to make rows of pictures. Have your child count by twos to tell how many pictures are in each row.

Leap Frog

Read the word on each frog. Write three words that rhyme with it.

1. 

2. 

3. 

4. 

5. 

6. 

Trace and write the sight words.

it

all

and

the

Choose a word family, such as *–at or –en*. Give your child a stack of index cards and have him or her write one word on each card that belongs with the word family. Examples: *-at— at, bat, cat, hat, fat* and *–en—Ben, den, hen, men.* Have your child use the cards like flashcards.

Hopping to 100

Fill in the chart to show all numbers from 1 to 100.

			4						
							18		
	22								
					36				
									50
		53							
61									
				75					
						87		89	

Help your child identify patterns in the chart. For example, if you start at 11 and move diagonally toward 99, you will notice that each number in the path has two identical digits. Also, for any number selected, you can move down one space to add 10 or up one space to subtract 10.

Read the story. Then answer the questions.

Logan's Toy Dog

Logan lost his toy dog. He looked everywhere. He looked on the chair. He looked on the table. He looked in the closet.

Logan cried, "It's lost! I will never find it." Then he saw a little ear sticking out from under the bed.

"There you are," said Logan. "I looked everywhere for you." He gave his toy dog the biggest hug ever.

1. Who is this story about?

2. Underline words in the story that help show where the boy is.
3. Where is he?

4. What is he looking for?

5. Where does he find it?

Have your child tell you a story about a time when he or she lost something. Encourage your child to compare that experience with the one shown in the story.

Before and After

Write the missing numbers.

1. 16 ___ 18

2. 33 ___ 35

3. ___ 24 ___

4. ___ 72 ___

5. 81 82 ___ 84 ___ 86

6. 13 14 ___ 16 ___ ___

7. ___ 62 63 ___ 65 ___

8. ___ ___ 41 42 ___ 44

Use the 100 Chart from page 26 or print a new one from the Internet. Cover some of the numbers on the chart with small squares of paper. Have your child tell you which numbers are covered.

Read the nursery rhyme. Then answer the questions.

Jack and Jill

Jack and Jill went up the hill
To fetch a pail of water;
Jack fell down and broke his crown,
And Jill came tumbling after.

1. Underline the words that rhyme in line 1 and line 3.
2. Why were Jack and Jill on the hill?
3. The rhyme says that Jack **broke his crown**.
   What do you think that means?
4. Practice reading the rhyme aloud. Read it at least three times.

Explain to your child that houses long ago did not have running water and that children often had the job of carrying water from a well to the house. Ask your child how life would be different today if we still had to collect water from a well.

On the Shore

Add.

1.

2 + 1 = ____

2.

3 + 4 = ____

3.

1 + 8 = ____

4.

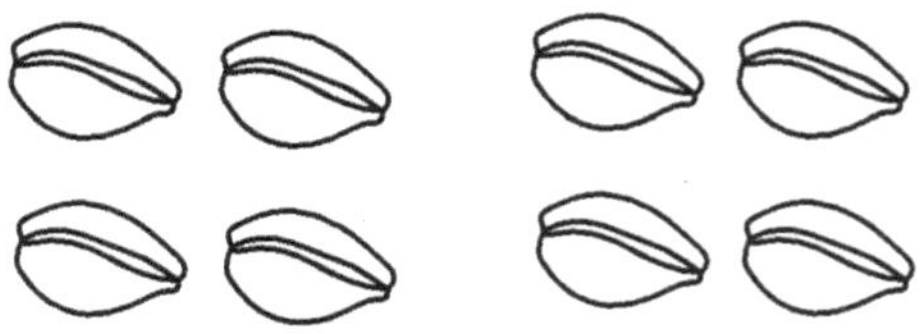

4 + 4 = ____

5.

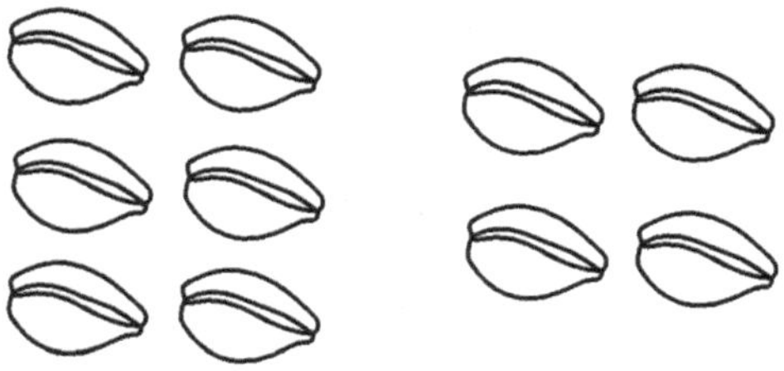

6 + 4 = ____

6.

3 + 2 = ____

7.

1 + 1 = ____

8.

4 + 0 = ____

Gather 10 counters for your child, such as beans, rocks, or beads. Tell your child a series of addition problems with sums of 10 or less. Have your child use the counters to show the problem and find the sum.

Listen to the Sounds

Write the beginning sound for each picture.

1. 

2. 

3. 

4. 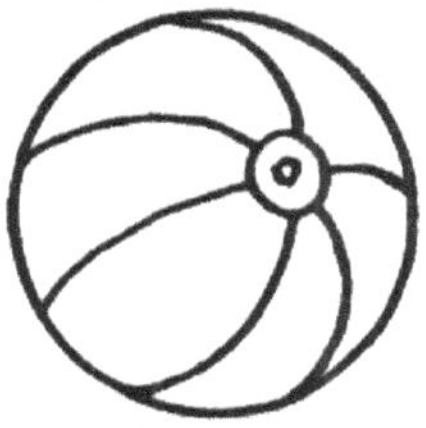

Write the ending sound for each picture.

5. 

6. 

7. 

8. 

Trace and write the sight words.

am

on

be

like

Draw a tic-tac-toe frame on a sheet of paper. Choose a commonly-used consonant for yourself and have your child do the same. Play tic-tac-toe using these two letters instead of the traditional x and o. Before writing a letter on the game board, players must first say a word that begins with that letter.

Smooth Sailing

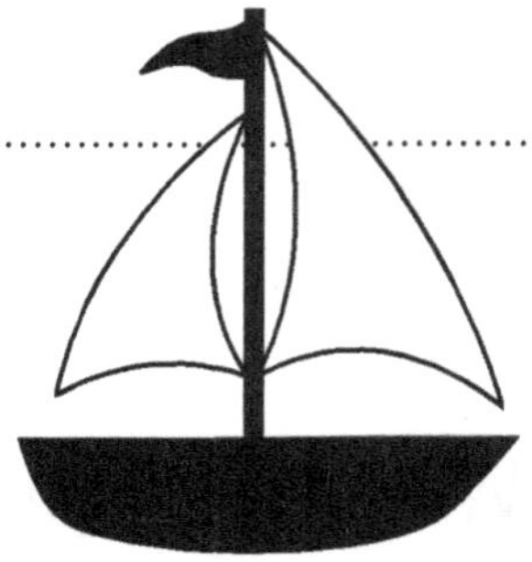

Draw a picture to solve each problem.

1.

8 − 2 = 6

2.

5 − 3 = ______

3.

6 − 4 = ______

4.

9 − 5 = ______

5.

10 − 7 = ______

6.

4 − 1 = ______

Help your child practice subtraction facts with a dice game. Each player rolls two dice and adds the numbers together. The player with the bigger number subtracts the other player's number and scores that many points.

Camp Mukaluk

Maps are tools that help people find places. Maps use symbols to help.

A compass rose is a symbol on a map. It shows north, south, east, and west.

A map key uses pictures to show where things are located on the map.

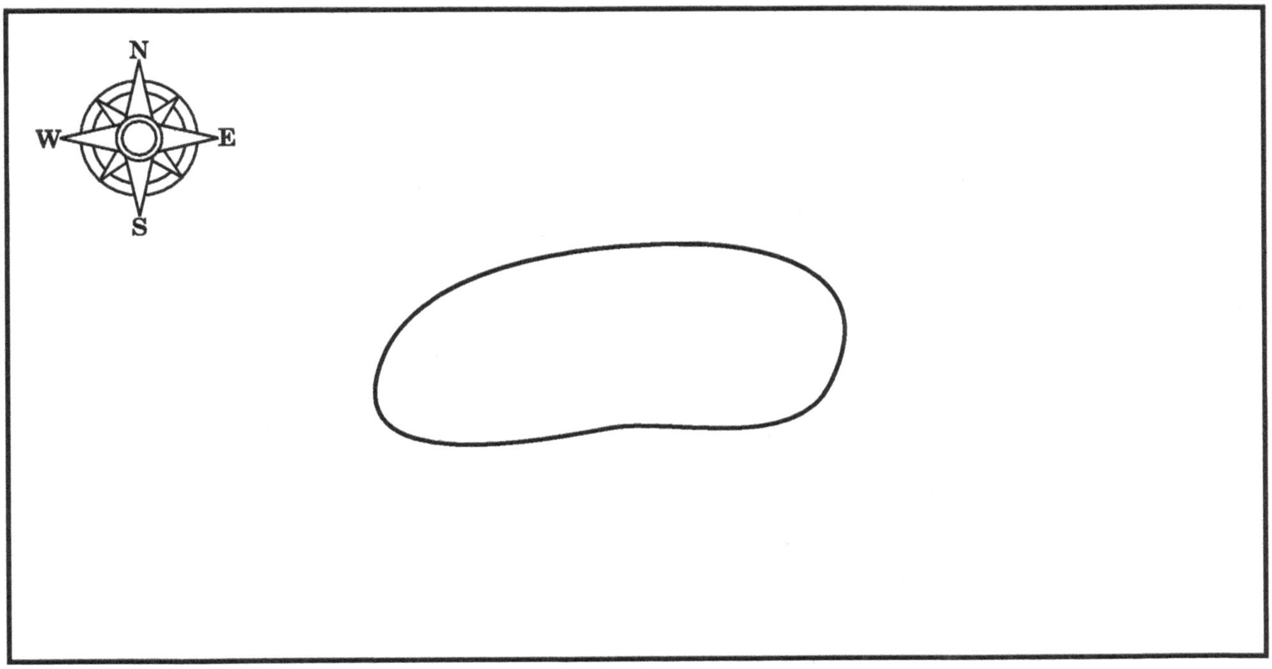

Follow the directions. Use the map key to show where things are at Camp Mukaluk.

1. Draw the Ranger Station north of the lake.
2. Draw the Picnic Area south of the lake.
3. Draw the Tent Camp east of the lake.
4. Draw the Forest west of the lake.
5. Draw the Restroom between the Tent Camp and the Picnic Area.
6. Draw a Trail around the lake.

Have your child practice making maps of places that he or she is familiar with. You might include a bed-room map, a park map, and a community map.

Adding Apples

Count the apples on the tree. Draw more to make 10.
Write the math sentence that matches the picture.

1.

□ + □ = □

2.

□ + □ = □

3.

□ + □ = □

4.

□ + □ = □

5.

□ + □ = □

6.

□ + □ = □

Use chalk to draw large, simple tree shapes on the sidewalk. Write a different number, 10 or smaller, on each tree. Have your child write addition problems on each tree with sums equal to the given number.

In the Middle

Write **a, e, i, o**, or **u** to spell each word.

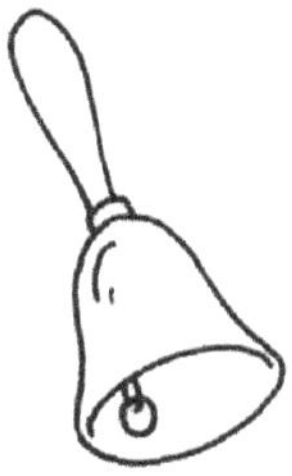

b ___ ll

d ___ ck

cr ___ b

f ___ x

fl ___ g

s ___ n

f ___ sh

c ___ t

Trace and write the sight words.

is

at

not

but

Make "Word Family Flip Books" for your child. Cut a strip of paper about six-inches long. Write the letter c on the left side and the word family *–at* on the right side. Cut several three-inch squares, stack them over the c, and staple them in place. Write a different letter on each square to make more words for the –at word family *(b, f, h, p, s, etc)*. Have your child flip through the book and read each word aloud. Make more books for other word families.

Number, Number, On the Line

Use a number line to solve each problem.

9 − 3 = 6

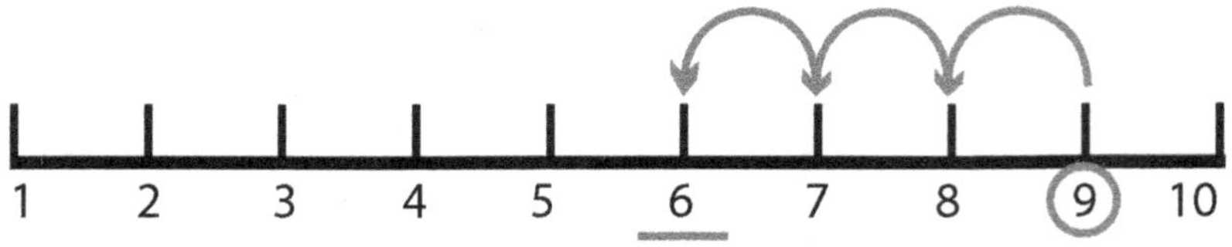

1. 10 − 4 =

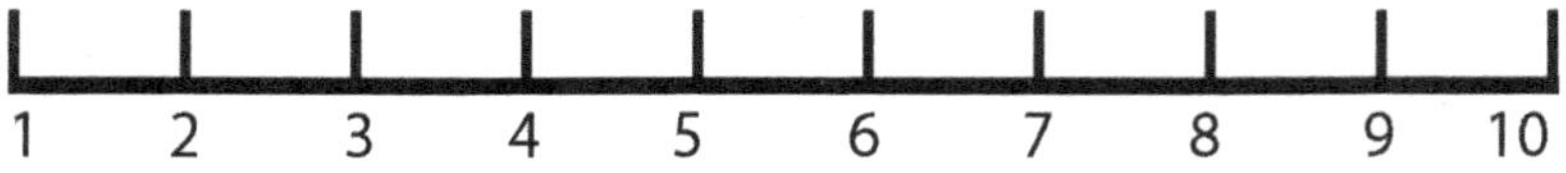

2. 5 − 1 =

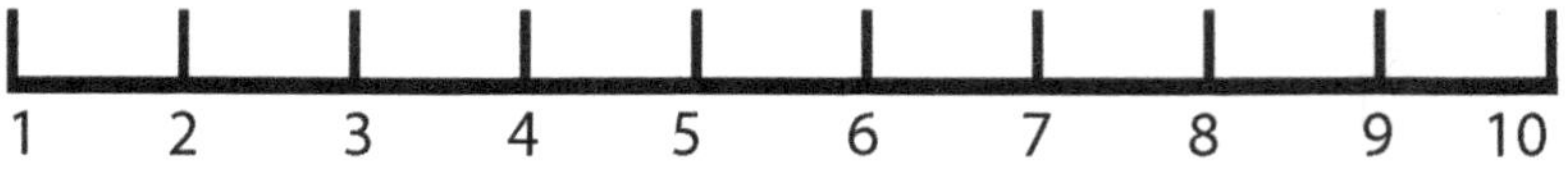

3. 8 − 3 =

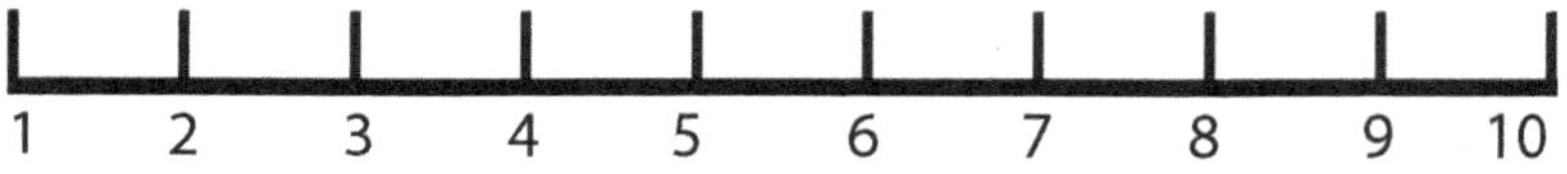

4. 9 − 5 =

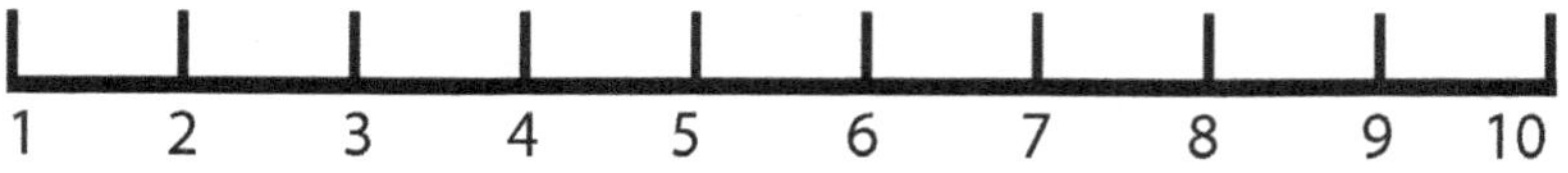

5. 6 − 2 =

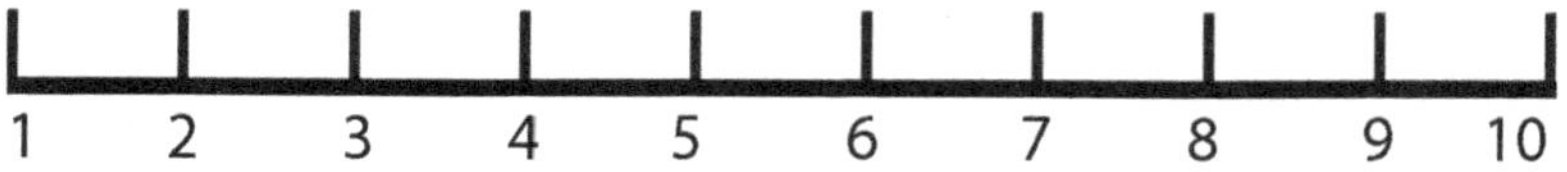

6. 7 − 6 =

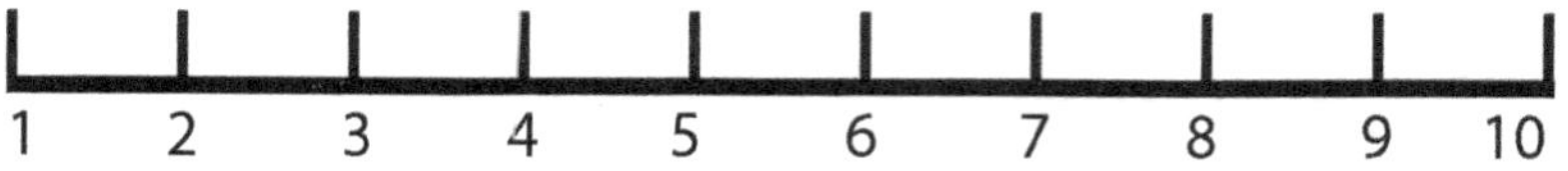

For this game, you need two teams with two players on each team. Use chalk to draw two hopscotch paths on the ground. Have one player from each team start at the top of the path on the number 10. The other player for each team rolls a single die and tells his or her teammate to hop that many squares down the path. The first team to hop to the end of the path wins the round.

Read the story. Then answer the questions.

Soccer

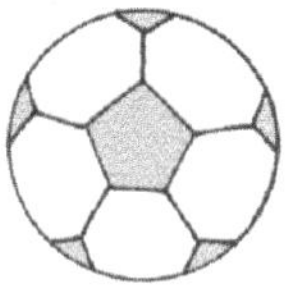

Ben plays soccer. He has a game on Saturday. Ben's sister likes to sing. The game will start at 9:30. Ben hopes to score a goal.

1. What sport does Ben play?

2. What time will his game start?

3. Cross out the sentence that does not belong in the story.

4. Draw a picture of Ben's game.

Have your child write a story about what they think happens at Ben's soccer game. Encourage him or her to use lots of sensory details.

School Days

Read the story. Draw a picture to solve the problem.

1. There are 3 cars on the rug. Kim adds 2 more. How many cars are on the rug in all? ______	2. Ted has 10 books in a basket. He takes out 3 books. How many books are left in the basket? ______
3. Jen makes a block tower with 10 blocks. She takes 4 blocks off. How many blocks are left in the tower? ______	4. Mark has 7 crayons. Kim gives him 3 more. How many crayons does Mark have in all? ______

Drawing a picture is one problem-solving strategy. Encourage your child to identify other strategies that can be used too, such as finding patterns, making tables or charts, and writing numbers in sequence. After solving word problems, always ask your child how he or she arrived at the answer.

Read the story. Then answer the questions.

Anna's Day at the Park

Anna went to the park. First, she went down the slide. Then, she dug in the sand. Next, she played on the swings.

"Time for lunch," called Anna's mom. Anna ran to eat. A day at the park was such fun!

1. Where was Anna?

2. What did she do first?

3. What did she do last?

Draw what you like to do at the park.

Help your child write a story that uses a sequence of events. Have your child use the *words first, then,* and *next* in the story.

Super Shapes

Count the number of shapes at the beginning of each row.
Color the set that has the same number of shapes.

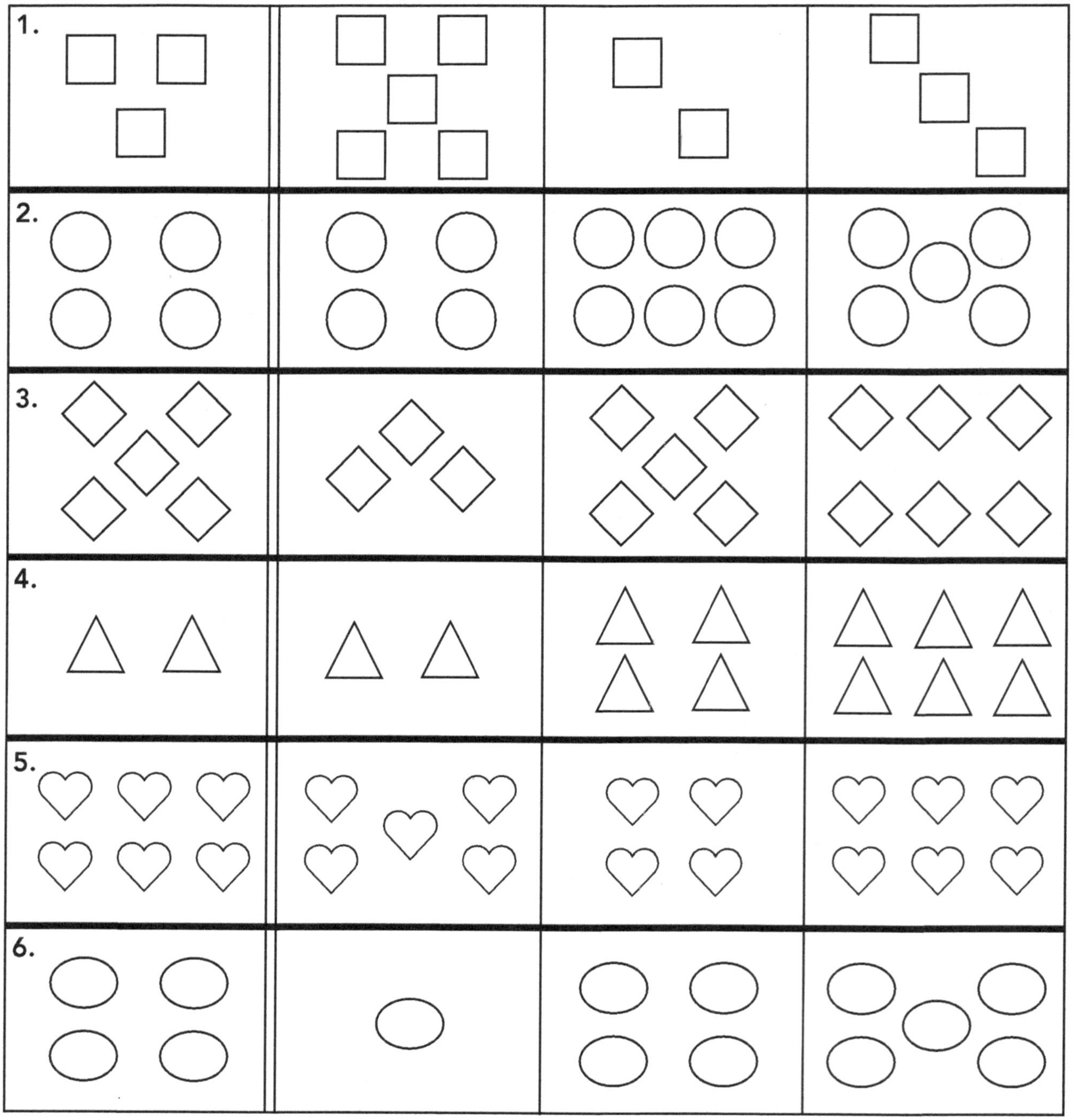

Help your child practice comparing numbers greater than 10. Glue sets of dry beans to index cards. Make two cards for each number you choose. Then mix up the cards and have your child find the matching pairs.

Cat's Park Day

Write the missing words.

down	for	in	she
to	will	go	

1. Cat went ________ the park.
2. She went ________ the slide.
3. She dug ________ the sand.
4. Where ________ ________ go now?
5. Cat will ________ home ________ lunch.

Write a sentence about what you would like to do at the park.

____

____

Write the sight words from this page on a set of index cards. Make a duplicate set. Shuffle the cards and arrange them face down in rows to play "Memory Match" with your child.

Fancy Flowers

Write the number of flowers in each set.
Color the set with the most flowers.

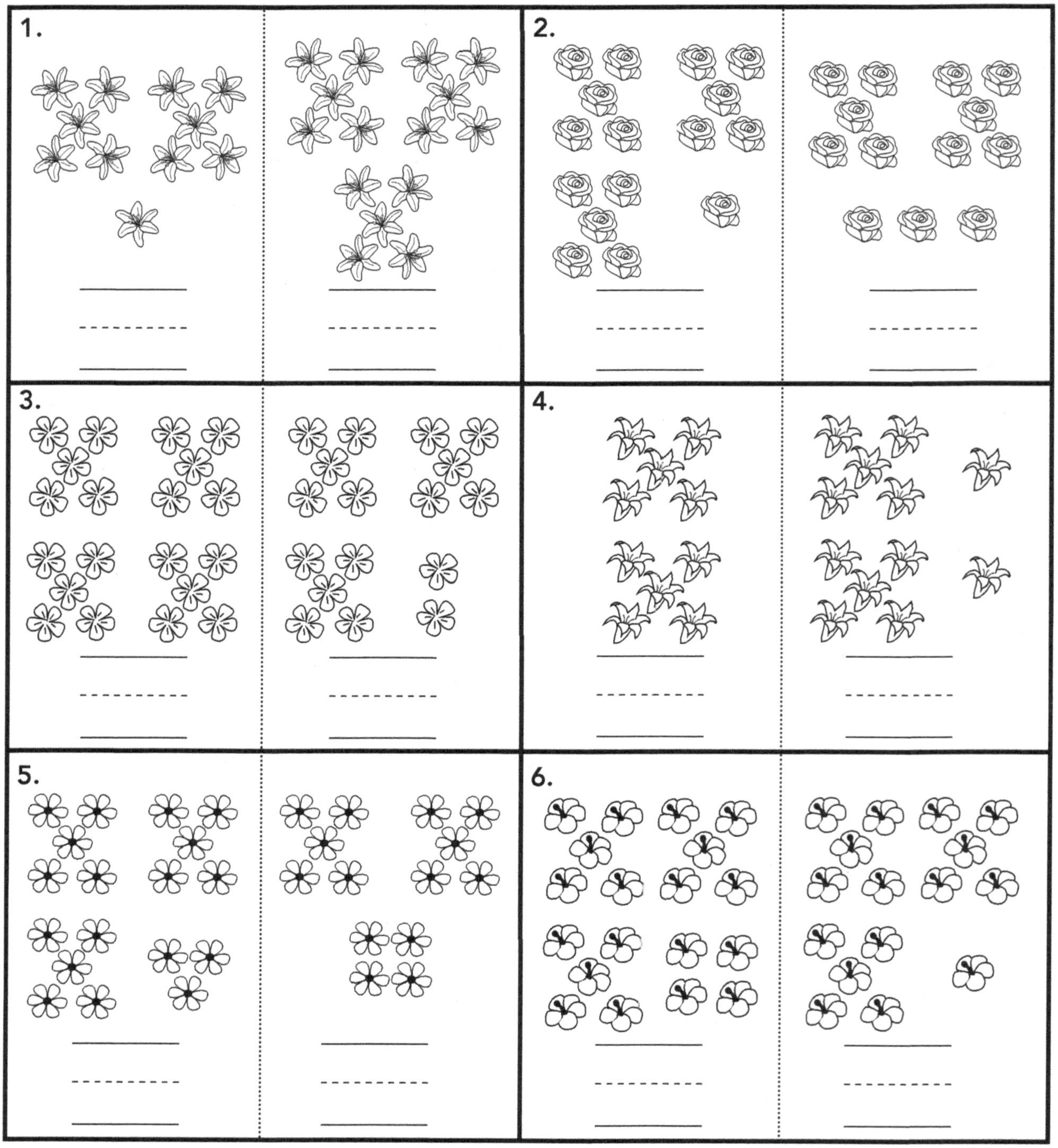

Name two numbers for your child and ask, "Which one is greater?" Then let your child roll out a tape measure and find the two numbers you named. Explain that the numbers on the tape measure get bigger as you move down the line.

Read the passage. Then answer the questions.

A Lake Habitat

A lake is a pool of water that has land all around it. Turtles, fish, and frogs live in lakes. Birds and bugs live there, too. Other animals live around lakes. They visit the lake to drink water and to look for food. Lakes are important habitats around the world.

1. What is a lake?

2. What are three animals that live in a lake?

3. Why would an animal visit a lake?

Plan a family trip to a lake or a pond in your neighborhood. Have your child bring a notebook and draw pictures of the plants and wildlife there.

Shape Up!

Count the shapes in each box. Write the number. Then draw a set of shapes that is less than the other set. Write the number for the set you draw.

1.

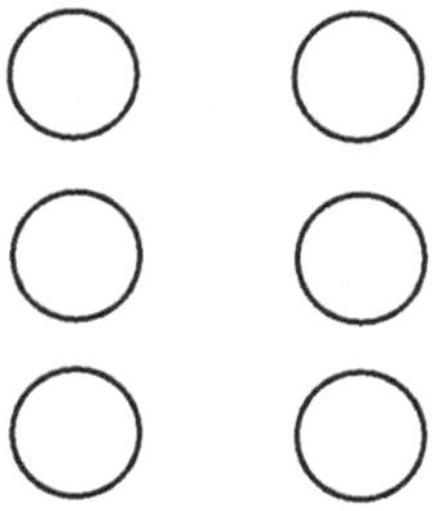

2.

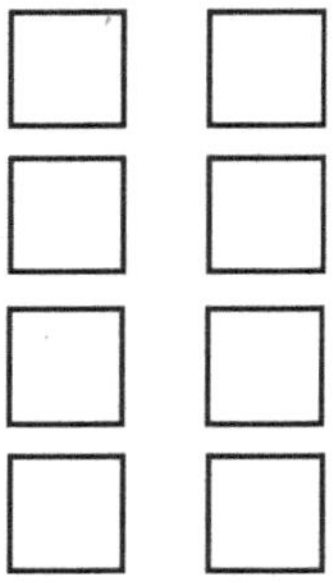

3.

4.

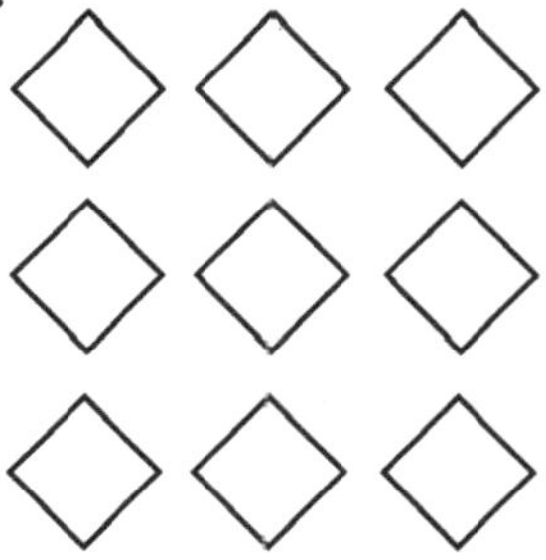

5.

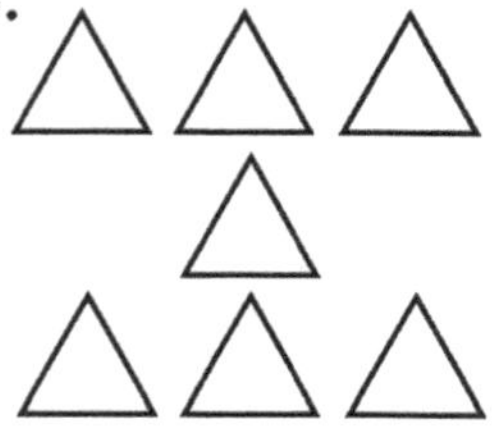

6.

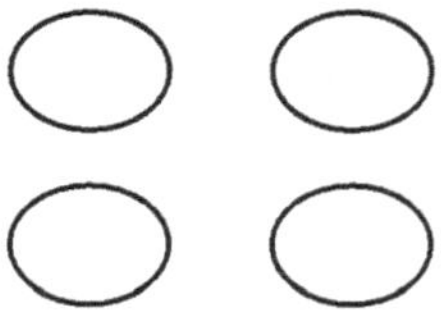

Have your child count the number of shoes present in each room of your house and record the data on a simple chart. Then have your child tell you which room has the least number of shoes in it and which room has the most. Continue with other objects that can be counted such as books, jackets, stuffed animals, etc.

Off to the Races

Circle the hidden words. Write them on the lines.

1. qandfg ________________ 2. rtdosm ________________

3. iareps ________________ 4. mwanty ________________

5. bseetr ________________

Use the words to finish the sentences.

6. I ran to ________________ Dad.

7. Dad said, "You ____________ fast!"

8. "Do you ________________ to race?" I asked.

9. "Yes, I ________________," said Dad.

10. We ran ____________ ran.

Have your child write a sentence for each word of the five words taught on this page.

Hungry Alligators

Circle the biggest number in each row.
Cross out the smallest number.

1.	**1**	**3**	**7**	2.	**11**	**13**	**15**

3.	**4**	**8**	**2**	4.	**17**	**10**	**14**
5.	**9**	**0**	**4**	6.	**19**	**20**	**18**
7.	**10**	**5**	**3**	8.	**17**	**19**	**14**
9.	**6**	**4**	**7**	10.	**12**	**16**	**18**
11.	**1**	**6**	**2**	12.	**16**	**9**	**12**

Set a timer for one minute and count how many times your child can hop on one foot. Record the number, and then have others take turns doing the same thing. Have your child identify who had the most hops. Try different activities as well, such as skipping, jumping jacks, and somersaults.

Read the story. Then answer the questions.

Surprise!

May and her dad went to the store. They bought a cake mix. They bought some candles. "Don't forget the frosting," said May.

May's dad added frosting to the basket. "Mom will be so surprised," he said.

1. Where did May and her dad go?

2. What are they buying?

3. What special event do you think is coming up?

4. Why do you think May's mom will be surprised?

Have your child draw a picture of the "perfect" birthday cake. Then have him or her help you bake a cake. Let your child frost the cake and add decorations.

Munch, Crunch

Use the code to color the picture.

Numbers GREATER THAN 50	Blue
Numbers LESS THAN 50	Green
Numbers EQUAL TO 50	Yellow

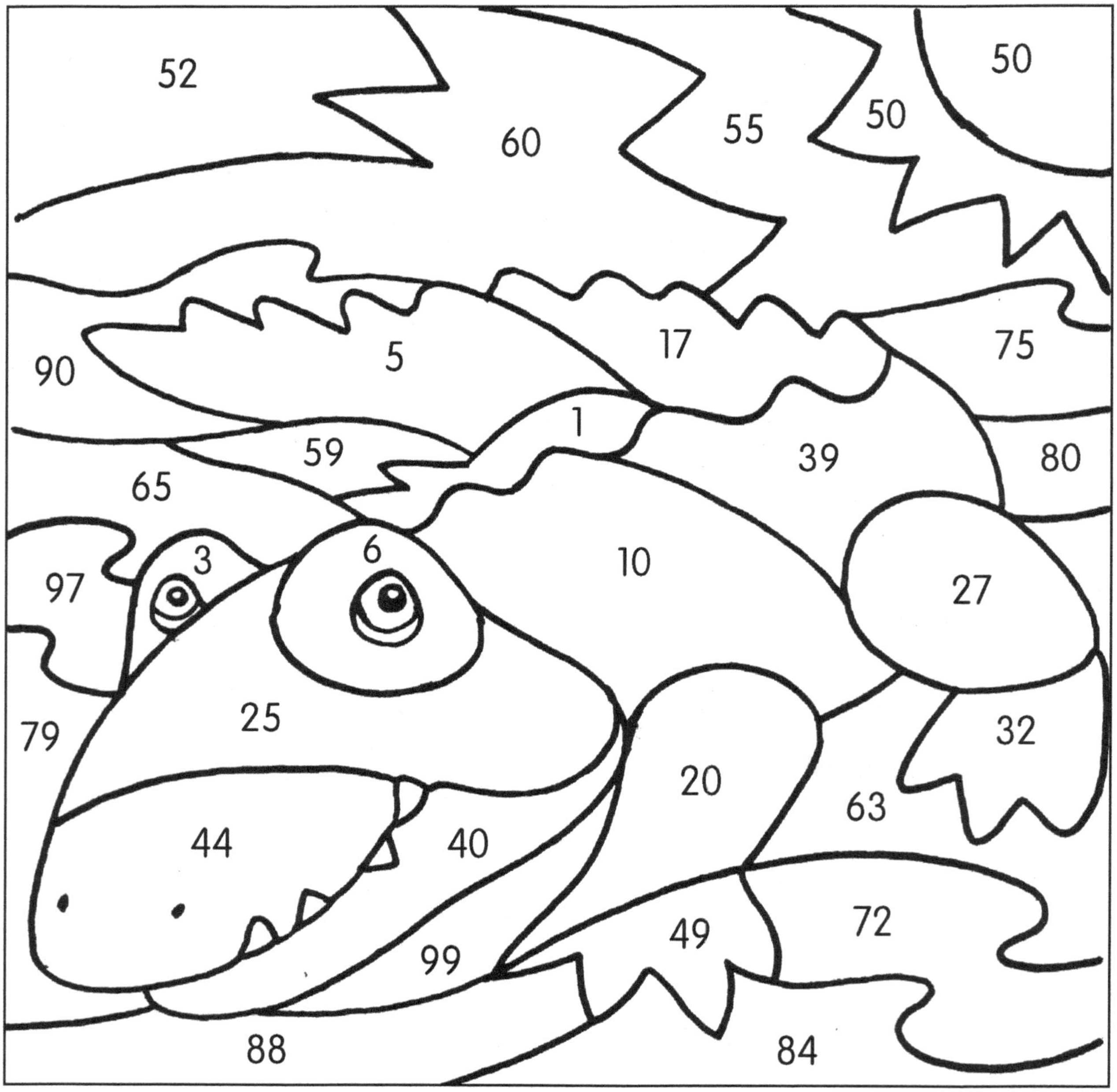

Use the hundreds chart on page 28, or print one from the Internet. Choose a secret number and have your child guess what it is based on "greater than/less than" clues. Then switch roles and have your child provide the clues while you guess the secret number.

Read the story. Then answer the questions.

Library Trip

Jenna went to the library. She saw a puppet show there. The show was funny. The puppets danced and told jokes.

After the show, Jenna found a book. It was about making puppets. She took the book home. "Now I can make my own puppets," said Jenna.

1. Where was Jenna?

2. What did she like about the puppet show?

3. What did the puppets do in the show?

4. What did she want to do after the puppet show?

Have your child write a short story about a small moment in their lives. Remind your child to include a beginning, a middle, and an end to the story.

Count the Change

Read the amount on each piggy bank. Write how many dimes and pennies you need to make that amount. Then write how many pennies you need to equal that amount.

	Amount	dimes	pennies	pennies
	19¢	1	9	19
1.	12¢			
2.	10¢			
3.	15¢			
4.	17¢			
5.	14¢			
6.	16¢			

Challenge your child to practice counting up to 50 pennies and trading sets of 10 pennies for 1 dime each.

All Around School

A noun names a person, place, or thing. Think of nouns you find at school. Write them on the chart.

Person	Place	Thing

Make a list of nouns that can be found in your neighborhood. Then take your child for a walk and see how many of the listed nouns he or she can find.

Circle Up

Circle 10. Count the extras. Add and write the number sentence. Then describe the group that is left.

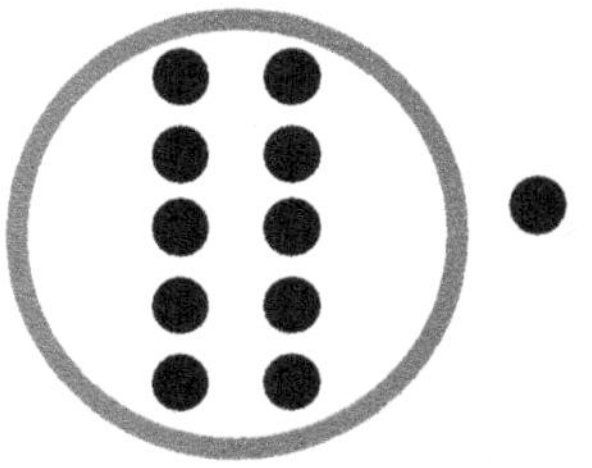

10 + 1 = 11

1.

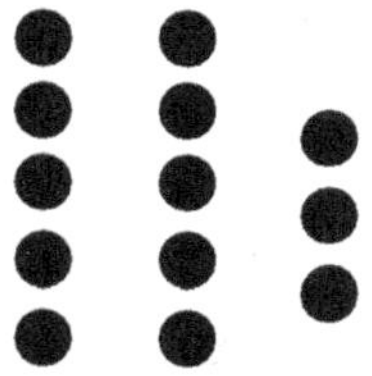

☐ + ☐ = ☐

2.

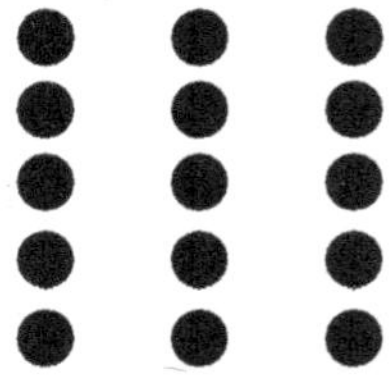

☐ + ☐ = ☐

3.

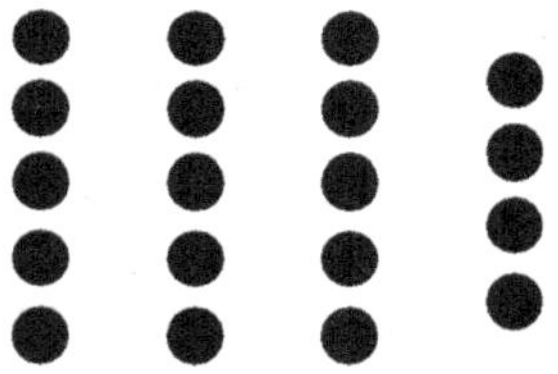

☐ + ☐ = ☐

4.

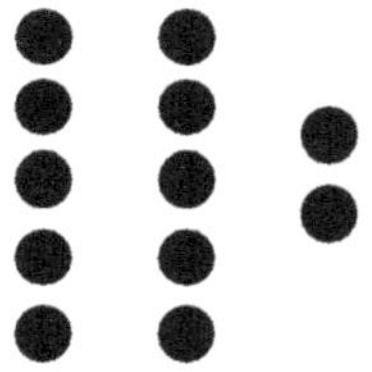

☐ + ☐ = ☐

5.

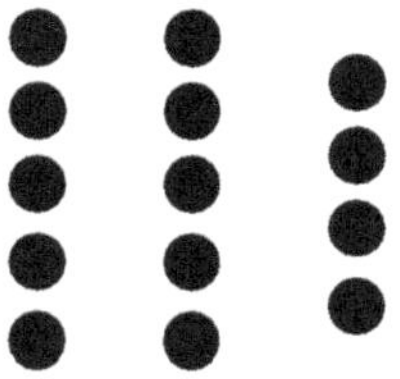

☐ + ☐ = ☐

6.

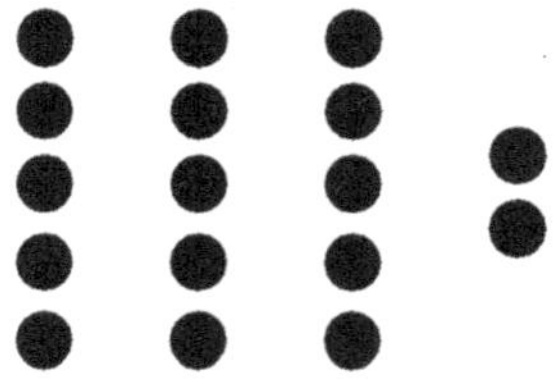

☐ + ☐ = ☐

Have your child string cereal loops onto lengths of yarn to make sets of 10. Then have him or her use single loops as "ones." Then have your child use the tens and ones to demonstrate different numbers. For example, 29 would be shown with two tens and nine ones.

Using a Timeline

Read the passage. Then answer the questions.

Sasha made a timeline in class.
It shows things she has done in her life.

Sasha's Timeline				
I was born.	I learned to walk.	I flew on a plane.	I rode a pony.	I started school.
2007	2008	2009	2010	2011

1. What year was Sasha born?

2. What happened the year before she started school?

3. Show the order of these events. Number them 1, 2, or 3.

____ Started school

____ Learned to walk

____ Flew on a plane

Give your child several photographs showing events in his or her life. Help place them in sequential order and glue them to a chart. Add the years in which each event happened to make a personal timeline for your child.

Criss-Cross

One thing in each row does not belong. Cross it out.
Then describe the group that is left.

1.

2.

3.

4.

5.

6.

   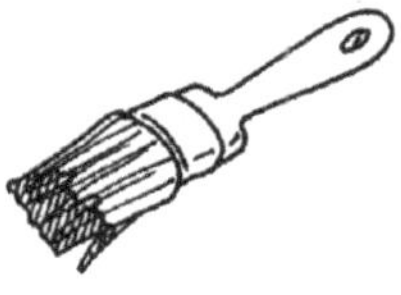

Cut basic shapes in different sizes from a variety of paper. Have your child sort the shapes and describe the "rule" for each group. For example, shapes can be sorted by size, color, number of corners, and so on.

Ready, Set, Action!

A verb is an action word. It tells what someone or something does. Use a gray crayon to color the parts of the picture that have verbs.

Have your child act out the movements of several different animals. Have him or her use a verb to tell what each movement is and then name other animals that may also move the same way. For example, animals that hop include rabbits, frogs, grasshoppers, and kangaroos.

What to Wear?

Color the things you wear in summer yellow.
Color the things you wear in winter blue.

Number of summer things ____________

Number of winter things ____________

Circle which has more. summer winter

Have your child classify different foods in your kitchen. Then have him or her compare the groups. For example, he or she may say, "There are more vegetables than fruits. There are fewer red foods than green foods."

Read the story. Then answer the questions.

The Giant Turnip—A Folktale

A farmer planted a turnip. It grew and grew. It grew so big, he needed help to pull it up.

The farmer called his wife. They pulled and pulled, but the turnip would not come up.

The farmer and his wife called their children. The family pulled and pulled, but the turnip would not come up.

The family called their pets. The farmer, his wife, their children, the dog, and the cat all pulled and pulled.

Pop! Out came the turnip.

The farmer chopped up the turnip. He cooked it into a great feast for everyone.

1. What did the farmer plant? ______________________

2. Why did he need help to pull it up? ______________________

3. Who helped the farmer? ______________________

4. Circle the word that tells what the farmer needed to pull up the turnip. teamwork anger bravery

Have your child tell you about a time that he needed to work with others to accomplish something. Add to the discussion with examples from your own life.

Tick-Tock

Write the time shown on the clock. Draw a picture of something you do at that time.

1

2

3

4.

5.

Help your child make a poster with hours listed from wake-up time to bedtime. Then add different activities that typically happen at different hours of the day.

Read the story. Then answer the questions.

At the Movies

Tomas and Ryan went to the movies. They sat in the front row. When the movie started, the lights went out. The music was loud. Tomas and Ryan watched the show. They loved it.

"Let's go again next week," said Tomas.

"Yes. Let's go again," said Ryan.

1. Where did Tomas and Ryan go?

2. What happened when the movie started?

3. What did the boys think of the movie?

Draw a picture about a movie that you have seen.

Have your child write a story about something he or she did with a friend. Encourage your child to include details about where they went, what they did, and how they felt about the experience.

Hot Stuff!

Add.

1. 3 + 4 =   5 + 2 =   9 + 3 =   7 + 5 =

2. 6 + 7 =   8 + 0 =   1 + 6 =   4 + 9 =

3. 10 + 10 =   2 + 8 =   0 + 12 =   13 + 1 =

4. 19 + 1 =   15 + 2 =   11 + 6 =   16 + 3 =

5. 17 + 2 =   20 + 0 =   18 + 1 =   14 + 6 =

Have your child practice solving addition problems by using a simple number line.

Yum-Yum

Adjectives are words that describe, or tell about, nouns.
Look at the picture on each cone. Write adjectives to describe it on the ice cream scoops.

Help your child practice using the parts of speech by teaching him or her to write cinquains. A cinquain is a five-line non-rhyming poem. It follows this format:

Line 1: A one-word noun
Line 2: Two adjectives that describe the noun
Line 3: Three verbs that describe the noun's actions and end in –ing
Line 4: A phrase that describes something about the noun
Line 5: A synonym for the noun

Kite Tails

Write three addition problems for the sum shown on each kite.

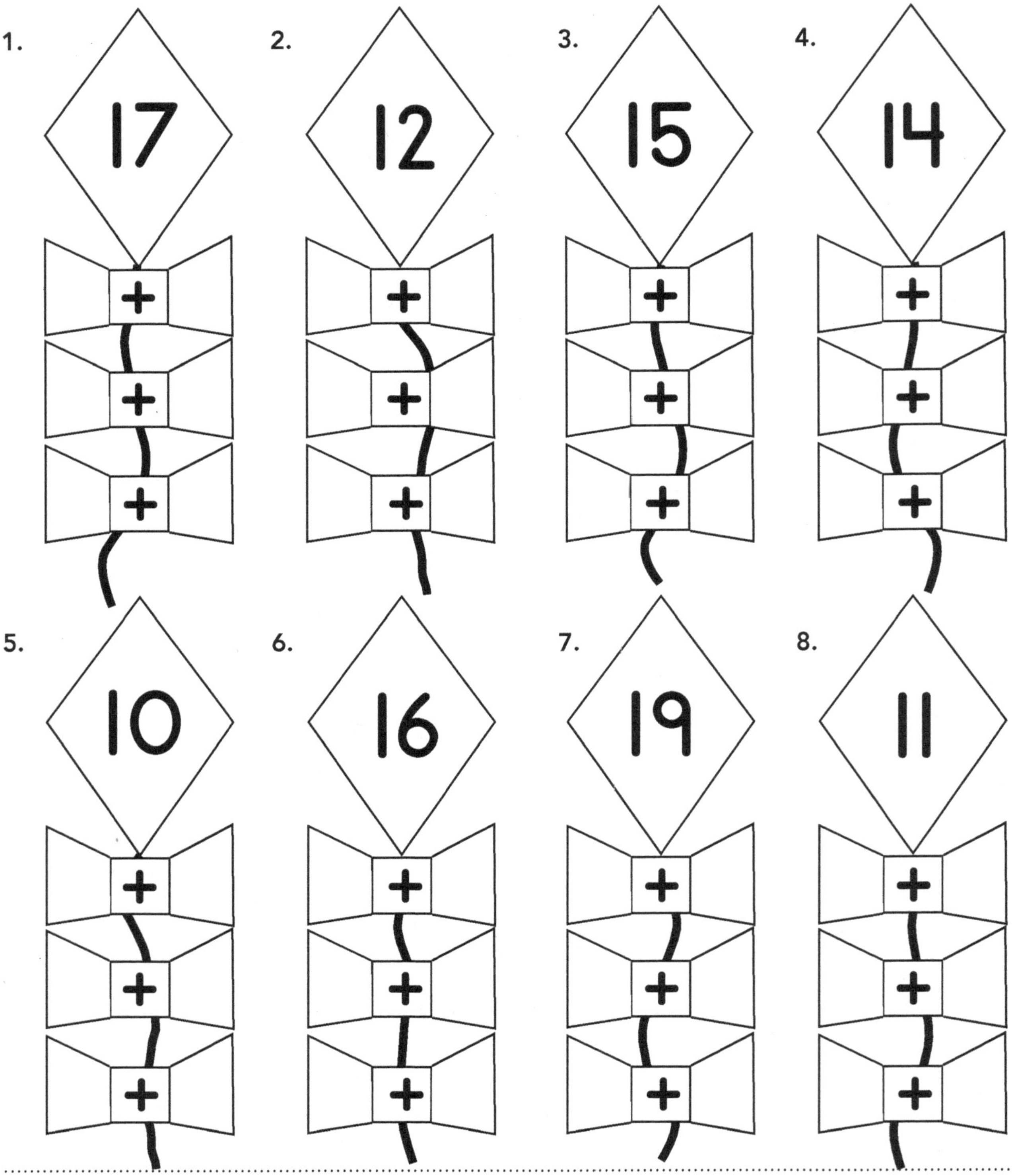

Choose a number less than 20 and write it on a sheet of paper. Then have your child list all of the pairs of numbers that can be added together to make that number. You may want to have your child start with + 0, then + 1, + 2, and so on.

Read the passage. Then answer the questions.

Reptiles

There are many kinds of reptiles. Snakes, lizards, turtles, and alligators are just a few. Reptiles lay eggs. Their bodies are covered with scales or hard, bumpy plates. They can not make their own bodies warm, so they lie in the sun when they need to warm up. Reptiles are interesting creatures.

1. What is this story about?

2. Name two kinds of reptiles.

3. Why do reptiles lie in the sun?

4. Underline the sentence in the story that tells you how the author feels about reptiles.

Have your child choose a reptile to learn more about. Help your child borrow library books about the animal and read them together.

Ten-Gallon Hats

Circle the numbers that make 10 in each problem.
Then add the third number.

1. 6, 4, 2 → 10 + 2 = 12

2. 7, 3, 6

3. 9, 2, 8

4. 4, 6, 1

5. 8, 2, 3

6. 5, 9, 1

7. 10, 0, 10

8. 1, 9, 7

9. 9, 3, 7

Have your child use stacking cubes to practice this skill. Give your child a three-digit addition problem to solve where two of the numbers add up to 10. Have your child use the cubes to count out all three quantities and then put the two sets together to make 10. Then have him or her add the third number to get the final answer.

Make a Match

Circle the sentence that is written correctly.

1. Rabbits hop.
   Rabbit hop.

5. She runs.
   She run.

2. I swim.
   I swims.

6. Cats naps.
   Cats nap.

3. We sings.
   We sing.

7. They sit.
   They sits.

4. He bake.
   He bakes.

8. You play.
   You plays.

Write a verb to complete each sentence.

I ______________________.

We ______________________.

Dogs ______________________.

It ______________________.

Have your child identify the verbs used on one page of a picture book. Then ask: "How would those verbs change if the nouns changed?" Give a few examples and then have your child offer his or her own examples.

Giddy-Up!

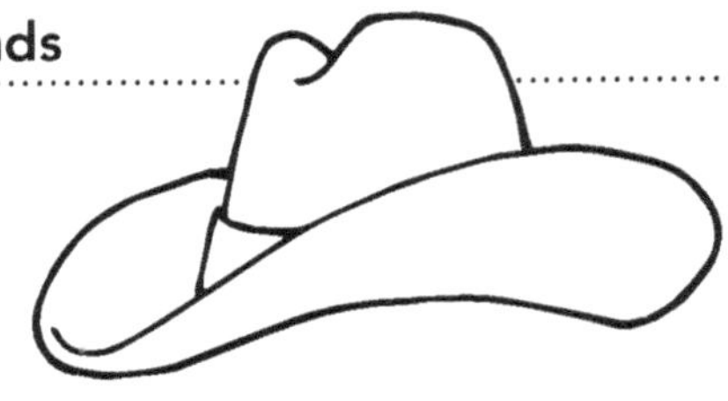

Fill in the missing number.

a 8 + □ = 11	b 9 + □ = 16	c 11 + □ = 15
d 0 + □ = 10	e □ + 9 = 20	i □ + 7 = 13
l □ + 12 = 14	o □ + 6 = 11	s 3 + □ = 15
t 16 + □ = 17	u □ + 8 = 16	w □ + 9 = 18

Write the letter for each number to solve the riddle.
Why did the horse sleep in the barn?

___ ___ ___ ___ ___ ___ ___   ___ ___   ___ ___ ___   ___ ___ ___
7 11 4 3 8 12 11   6 1   9 3 12   1 5 5

___ ___ ___ ___   ___ ___ ___ ___ ___ ___ ___ !
4 5 2 1   5 8 1 12 6 10 11

Have your child practice the math strategy of counting on from a given number to reach a sum. For example, for 8+□=11 have your child start at 8 and count how many numbers it takes to move to 11 (3).

Read the story. Then answer the questions.

Blake's Adventure

Blake looked around. No one could see what he was up to. He crept out of the garden. He lifted the big gray stone by the gate. A secret door appeared in the wall. Blake pushed it open and walked through. A sparkling green light bounced over the path in front of him. Blake followed.

1. Is this a story about something real or pretend? How do you know?

2. Write a word that describes Blake.

3. Write what you think happens next?

Divide a sheet of white paper into a 3 x 3 grid. Label the rows *Beginning, Middle,* and *Ending*. Have your child draw a picture of three things Blake did at the beginning of this story. Then ask your child to imagine what happens in the middle and end of the story. Have him or her draw three pictures for each section and write a sentence for each one.

Down on the Farm

Solve each problem. Draw a picture or write a sentence to explain your answer in the box.

1. Sue put 6 eggs in the basket. Tim put 5 eggs in the basket. How many eggs were in the basket?	2. Ashton grew two rows of carrots. The first row had 9 carrots. The second row had 4 carrots. How many carrots did Ashton grow?
3. Madison counted 3 sheep in the barn. Then 4 more sheep came in. How many sheep were in the barn then?	4. There were 6 cows by the tree, 4 cows by the gate, and 7 cows in the barn. How many cows were there in all?
5. There were 5 pumpkins in the house, 5 pumpkins on the porch, and 3 pumpkins in the garden. How many pumpkins were there in all?	6. The farmer had 7 horses. He bought 3 more on Friday. Then he bought 10 more on Saturday. How many horses does the farmer have now?

Create story problems for your child to solve that call for the addition of two or three numbers. The solution for each addition problem should be 20 or less.

Read the poem. Then answer the questions.

The Animal Show

On your marks! Get set! Let's go!
We're going to see the animal show.
What will we see once we are there?
Animals common and animals rare.
Animals meek and animals wild.
Animals to thrill any child.
Loud ones, quiet ones, long and short.
Animals that bark, roar, and snort.
We're off to see the animal show.
Get ready. Get set. It's time to go!

1. Circle the word that describes the author's feelings about going to the animal show.

   scared   excited   sad

Underline words in the poem that help create that feeling.

2. What do you think the word **meek** means?

3. Think about the words used to describe the animals. Write three animals that might be at the show.

Have your child draw a picture of his or her favorite animal and write a sentence that tells something about it.

Surf Safari

Subtract.

1. 7 − 0 = 10 − 1 = 4 − 1 =

2. 8 − 2 = 3 − 3 = 11 − 3 =

3. 14 − 4 = 15 − 5 = 9 − 4 =

4. 9 − 6 = 16 − 6 = 12 − 7 =

5. 20 − 10 = 17 − 8 = 13 − 8 =

6.

12	9	6	13	8
− 0	− 1	− 2	− 2	− 5

7.

20	18	11	19	15
− 5	− 6	− 7	− 9	− 10

Play hopscotch with your child. Before each turn, say a subtraction problem and have your child solve it.

Crazy for Capitals

Use a capital for:

The first word in a sentence
I
Names of the months and days
Names of people and places

Circle each letter that needs a capital.

POSTAGE

august 2, 2013

Dear zoe,

i am on a trip with my family. it is lots of fun. on monday, we swam at green lake. on tuesday, we hiked to bear falls. tomorrow i will ride a horse.

we will be home soon. i can't wait to see you.

Love,

amanda

Have your child write a letter to a friend that tells about something he or she has done this summer.

Under the Sea

Subtract. Use the code to color the picture.

1—5 Blue	11—15 Yellow
6—10 Green	16—20 Red

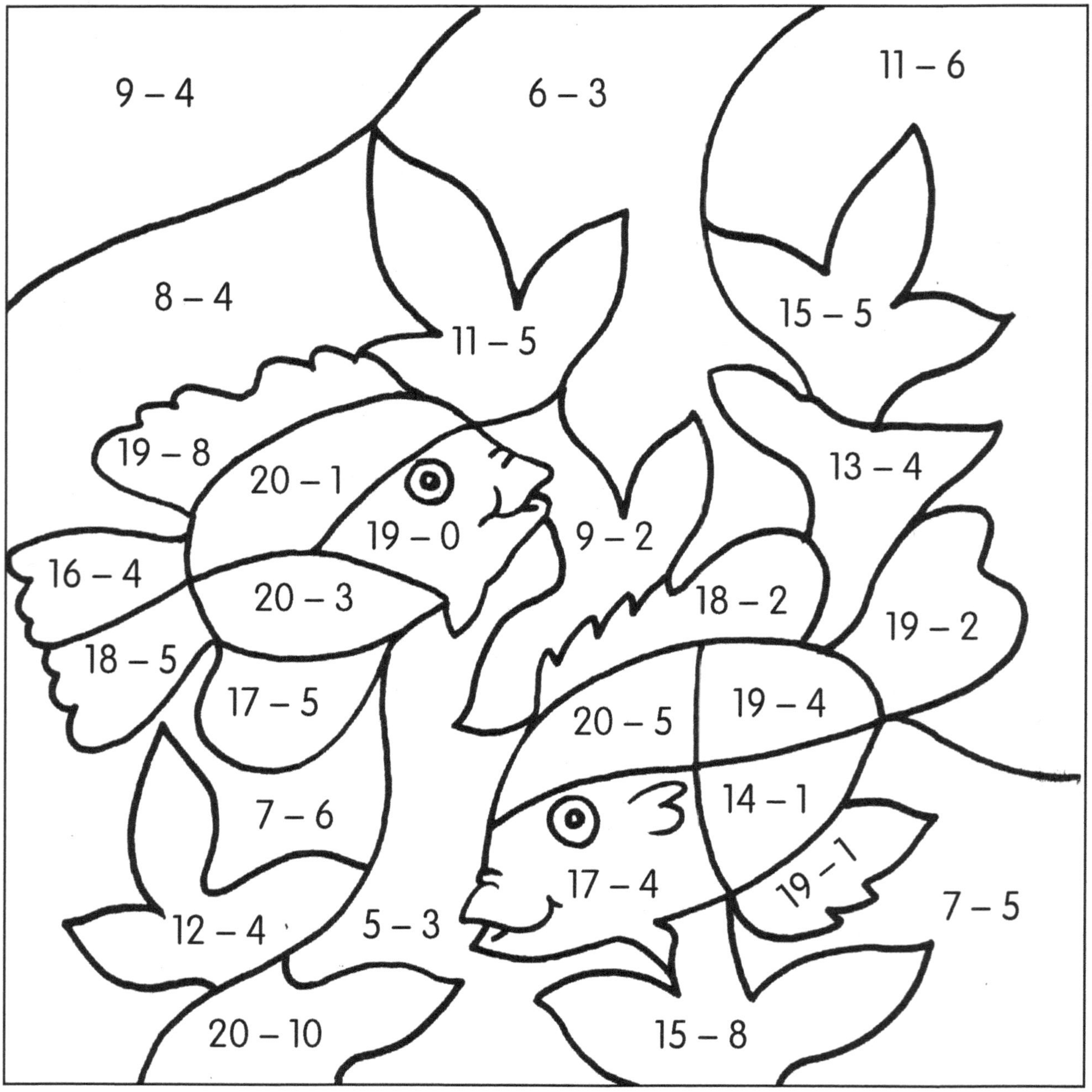

Collect 24 craft sticks. Number the sticks from 0 to 20. On the remaining sticks write +, –, and =. Have your child arrange the sticks to make true number sentences, such as 6 + 11 = 17 and 15 – 8 = 7.

Using a Table of Contents

Informational books tell us facts about a topic. A table of contents is a page at the beginning of an informational book. It lists the parts of the book. It gives the page number that begins each part.

Read the Table of Contents. Then answer the questions.

Africa

Table of Contents

1.What is this book about?

2.On what page would you find a map?

3. If you wanted to read about animals that lived in water, what part would you read?

4.What page does "Life in the Deserts" start on?

5.What part of the book would tell you the meanings of words?

Ask your child what things he or she knows a lot about. Have him or her choose a topic and write a simple book about it. Have your child write a table of contents to go with the book.

A Whale of a Tail

Write the number that completes the math sentences on the whale's tail.

1. ☐
   $10 - \square = 8$
   $8 + \square = 10$

2. ☐
   $9 - \square = 6$
   $6 + \square = 9$

3. ☐
   $7 - \square = 5$
   $5 + \square = 7$

4. ☐
   $4 - \square = 3$
   $3 + \square = 4$

5. ☐
   $8 - \square = 3$
   $3 + \square = 8$

6. ☐
   $5 - \square = 1$
   $1 + \square = 5$

7. ☐
   $10 - \square = 4$
   $4 + \square = 10$

8. ☐
   $9 - \square = 2$
   $2 + \square = 9$

For each problem shown here, challenge your child to complete the fact family. Have him or her tell you what the remaining two math sentences are for each set of three numbers.

Beach Day

All sentences end with a punctuation mark.

.	period	use at the end of a telling sentence
?	question mark	use at the end of an asking sentence
!	exclamation mark	use at the end of a sentence with strong feelings or a surprise

Read each sentence. Write the correct punctuation mark for each one.

1. I went to the beach
2. Can you make a sand castle
3. Watch out for the waves
4. The sun is hot
5. Let's race to the water
6. This is so much fun
7. I see a shell
8. Can we come back again

9. Write a telling sentence.

10. Write a question.

11. Write a sentence with a strong feeling.

Play this version of "Mother, May I?" with your child. Stand at one end of the play area with your child facing you at the other end. Say a sentence or a question. For each sentence, have your child take two steps forward. For each question, have him or her take one step backward. The game is over when your child reaches your side of the play area.

Dolphin Dash

Count by twos to subtract. Use the number line to help.

18 - 6 = 12

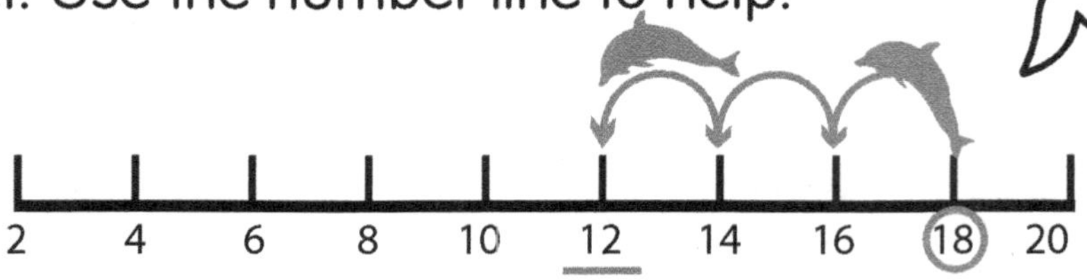

1. 10 - 4 = ☐ 2 4 6 8 10 12 14 16 18 20
2. 20 - 6 = ☐ 2 4 6 8 10 12 14 16 18 20
3. 18 - 2 = ☐ 2 4 6 8 10 12 14 16 18 20
4. 14 - 10 = ☐ 2 4 6 8 10 12 14 16 18 20
5. 16 - 8 = ☐ 2 4 6 8 10 12 14 16 18 20
6. 18 - 8 = ☐ 2 4 6 8 10 12 14 16 18 20
7. 20 - 8 = ☐ 2 4 6 8 10 12 14 16 18 20
8. 10 - 6 = ☐ 2 4 6 8 10 12 14 16 18 20

Use chalk to write the numbers 0 to 20 on the sidewalk. Say a subtraction problem for your child and have him or her hop through the number line to find the answer.

Read the fable. Then answer the questions.

The Dog and His Shadow

It was market day in the town. People set up tables to sell all sorts of things. There were carrots and cabbages, socks and blankets, eggs and chickens.

A hungry dog walked by the tables. He sniffed the air and hoped for a treat. A kind old man saw the dog and gave him a nice bone.

The dog wanted the bone all to himself, so he set out for home. On his way, he crossed a bridge over the river. He looked into the water and saw his own reflection looking back.

"There's another dog," he thought. "He has a bone bigger than mine. I am so hungry, I must have both bones."

At that the dog jumped into the river. His reflection disappeared in all the splashing and the bone was lost for good deep in the river.

Retold from Aesop's Fables

Enjoy what you have instead of chasing after what isn't there.

1. Who gave the dog a bone?

2. Why did the dog want to go home?

3. What did the dog see in the river?

Point out the moral of the story: *Enjoy what you have instead of chasing after what isn't there.* Point out that many fables try to teach important life lessons. Brainstorm a few life lessons with your child and write a fable together to teach one of the lessons on your list.

Gone Fishing

Solve each problem. Draw a picture or write a sentence to explain your answer in the box.

1. Gus has 10 fishing boats. He sent 5 boats out on trips. How many boats did not go out?	2. 8 dolphins swam next to a boat. 4 dolphins swam away. How many dolphins stayed by the boat?
3. Captain Jack made 18 pancakes for his crew. The crew ate 12 pancakes. How many pancakes were left?	4. There were 19 seagulls on the dock. First, 4 flew away. Then 5 more flew away. How many seagulls were still on the dock?
5. There were 14 fishing poles on the boat. Jane took 3 poles. Tom took 7 poles. How many poles were left?	6. Tina caught 20 fish. She gave 4 to Robin. Then she gave 9 to Erin. How many fish did Tina have left?

Create story problems for your child to solve that call for the subtraction of two or three numbers from a quantity of 20 or less.

Read the story. Then answer the questions.

Flower Fairy

There was a shiny speck on a flower. Brenna looked closer. The shiny speck moved. It hopped from flower to flower. Brenna looked even closer. The speck had wings and a tiny dress! Brenna had found a flower fairy.

1. Who saw the shiny speck?

2. What was it doing?

3. What was on the speck?

4. What had Brenna found?

5. Write what you think happens next.

Have your child write a story from a fairy's point of view.

Buzz Buzz

Count by ones and write the missing numbers.

86 ____ ____ ____ ____ ____ ____ 93

54 ____ ____ ____ ____ ____ ____ 61

98 ____ ____ ____ ____ ____ ____ 105

Count by fives and write the missing numbers.

60 ____ ____ ____ ____ ____ ____ 95

75 ____ ____ ____ ____ ____ ____ 110

85 ____ ____ ____ ____ ____ ____ 120

Count by tens and write the missing numbers.

50 ____ ____ ____ ____ ____ ____ 120

10 ____ ____ ____ ____ ____ ____ 80

40 ____ ____ ____ ____ ____ ____ 110

When doing laundry, have your child help find matching socks and then count by twos to tell how many socks there are.

Just Add E

Read the words. Write **e** at the end to change the short-vowel sound to a long-vowel sound.

fin______ tot______ cut______ Sam______

Choose the word to complete each sentence. Circle the word with the long-vowel sound.

1. I ______ skip. She ate a candy ______ .

can cane

2. See the ______ fly. I have a model car ______ .

kit kite

3. I ______ to see you soon. The rabbit can ______ away.

hop hope

4. The ______ took a nap. A block is a ______ .

cub cube

Use the short-vowel and long-vowel words on this page for a memory match game. Write each word on a separate index card. Shuffle the cards, then place them face-down in rows. Have your child turn over two cards at a time to find the pairs that show the short-vowel word and the same word with the –e added at the end.

Hippity-Hop

Write the number that comes before.

____, 101 ____, 93 ____, 120

____, 87 ____, 111 ____, 70

Write the number that comes after.

108, ____ 102, ____ 118, ____

115, ____ 78, ____ 54, ____

Find the pattern. Write the missing numbers.

93, 96, ____, 102, 105, ____, 111

106, 108, 110, ____, ____, 116, ____

____, 95, 100, ____, ____, 115, 120

Use a deck of playing cards for this game. Shuffle the number cards and aces (for ones). Deal three cards to each player. Each player makes the largest two-digit number possible from the cards in their hands and places the cards face-up. The player with the largest number tells what number comes before and after each number that has been created. Then he or she keeps all of the cards. To continue playing, deal two more cards to each player. The player with the most cards at the end wins.

Read the passage.

Seasons

Earth takes one year to travel around, or **orbit**, the sun. During this orbit, Earth tilts and spins. The part of Earth that leans toward the sun has summer. The part that leans away has winter. The seasons change around the world as Earth continues its orbit.

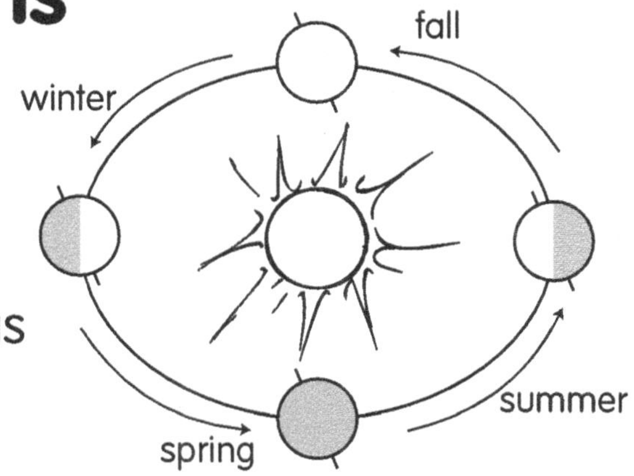

Read each sentence. Write **true** or **false**.

1. It takes one year for Earth to orbit the sun. ____________

2. Summer happens when Earth leans away from the sun. ____________

3. Seasons change as Earth orbits the sun. ____________

Draw a picture of the clothes you wear in summer and winter.

Summer	Winter

The dates marking the first day of each season change from year to year, but they generally fall around the 20th for the months of March, June, September, and December. Use an almanac to get the dates for this year and help your child find and mark the dates on a calendar.

Flower Power

Count the tens and ones. Write the number.

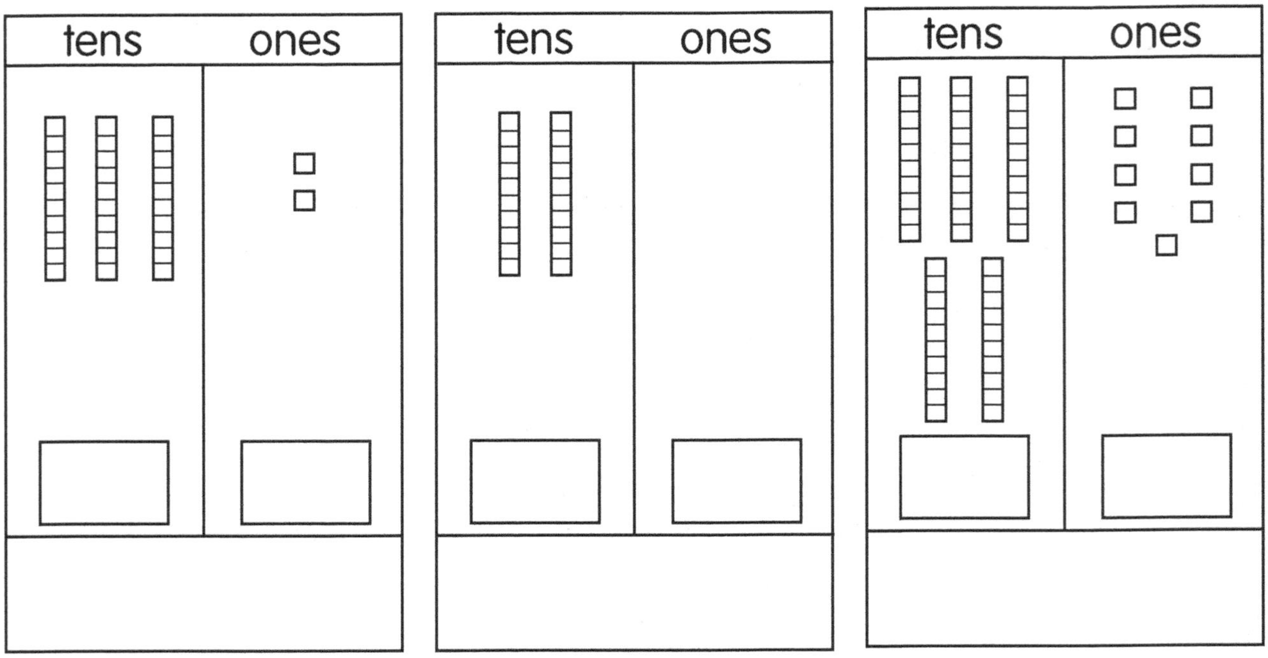

Draw tens and ones to show the number.

tens	ones	tens	ones	tens	ones
17		40		35	

Cut 10 strips of paper and have your child put 10 thumbprints on each one. Then roll a paper grocery bag into a long strip and staple it at the ends to make a hatband. Staple the paper strips to the hatband to make a "hat of 100."

Make the A Say A

There are different letter combinations that make the **long a** sound.

final-e	ai	ay
make	tail	day
lane	rail	say

Circle the correct spelling of each word.
Write the word in the sentence.

1. I got a letter in the _______________ . mail mayl mal

2. Ella will _______________ with Ethan. staye stay stai

3. Ashton was _______________ to school. layt lait late

4. Olivia broke the _______________ crayon. grai graye gray

5. Marcus has a _______________ of shells. pale pail payl

6. It will _______________ today. rain rane rayn

Have your child write a story that includes some of the *long* a words shown on this page.

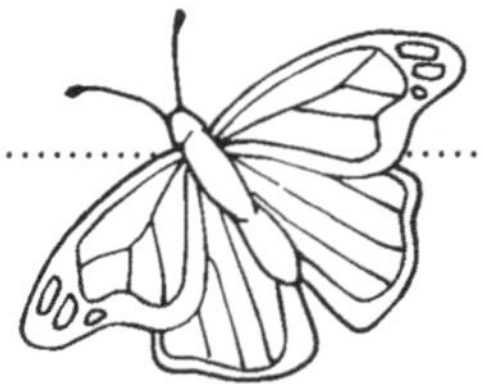

Garden Beauties

Look at the tens first. The number with more tens is greater than the other number.

58 > 32

23 < 49

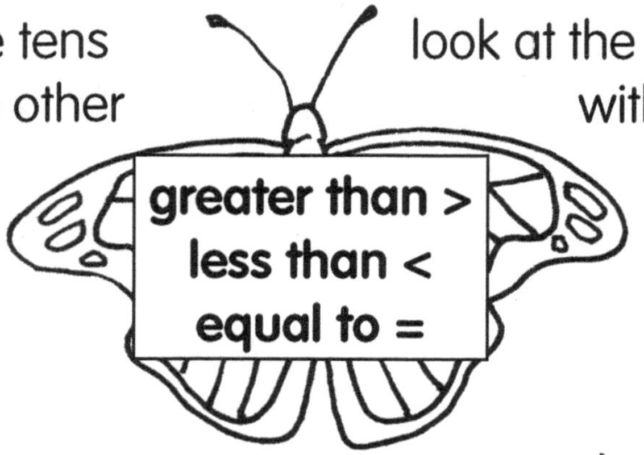

If the tens are equal, then look at the ones. The number with more ones is the greater number.

64 > 61

92 < 95

Write >, <, or =.

97 ○ 43   22 ○ 36   52 ○ 57

61 ○ 61   18 ○ 11   74 ○ 47

28 ○ 84   40 ○ 60   59 ○ 51

13 ○ 33   44 ○ 24   38 ○ 17

99 ○ 79   82 ○ 82   65 ○ 77

Practice comparing numbers with your child by playing this fun guessing game. Write a number less than 100 on a sheet of paper, but don't let your child see it. Have him or her guess what number you chose. Then tell your child that your number is greater than or less than the number guessed. Keep playing until your child guesses the correct number and then switch roles.

Read the story. Then answer the questions.

Time for School

It was the first day of school. Hannah sat on a hard bench between her two sisters. She had never been in a school before. She learned at home. Now she had a teacher—Miss Green.

Miss Green used a stick to write letters in the dirt floor. She called on students to name the letters. The students who did well moved to the front row. Hannah was proud that she and her sisters got to sit up front. She looked for her friend, Ruth. Poor, Ruth. She was in the back.

At lunch, Hannah and Ruth stood in line at the water bucket. They each filled a small cup. Then they sat under a tree to eat. "Don't worry, Ruth," said Hannah. "You will learn your letters. I will help you."

1. Does this story take place today or long ago? ______________

2. Which detail does not help tell when the story happened?

   The school had a dirt floor. It was the first day of school.

   The girls got water from a bucket.

3. Why does Hannah sit in the front of the room?

____

4. How do you think Hannah will help Ruth?

____

Have your child compare and contrast schools today with schools from long ago.

Oh, Nuts!

When you add 10 to a number, the tens column increases by 1. The ones column stays the same.

55 + 10 = 65

Add 10.

1. 23 + 10 = ☐　　14 + 10 = ☐　　39 + 10 = ☐

2. 40 + 10 = ☐　　76 + 10 = ☐　　81 + 10 = ☐

When you subtract 10 from a number, the tens column decreases by 1. The ones column stays the same.

97 - 10 = 87

Subtract 10.

3. 52 − 10 = ☐　　68 − 10 = ☐　　94 − 10 = ☐

4. 26 − 10 = ☐　　41 − 10 = ☐　　37 − 10 = ☐

Have your child roll two dice and create a two-digit number from the numbers rolled. Then have him or her add 10. After several turns, switch to subtracting 10.

Listen to the story. Then answer the questions.

The Frog Prince

Once upon a time, a princess went out to play. She tossed a golden ball up and down, again and again. At last the ball fell into a deep well. The princess cried and cried.

A slimy frog heard the princess. He said, "Promise to be my friend and I will dive into the well to get your ball." The princess promised, but as soon as she had her ball again, she skipped off without the frog.

That night the frog came to the castle. He told the king about the promise the princess had made.

"My dear girl," said the king. "If you made a promise, you must keep it." Then the king invited the frog to stay for dinner.

The princess was sad. She did not want a frog for a friend, but the frog came every night. He shared a plate and a cup with the princess. He even slept on her pillow.

One morning, as the frog hopped off the pillow, something magical happened. The frog turned into a prince. "I was under a spell," he told the princess. "I needed a friend to become a boy again."

The prince and the princess ran off to play. They have been friends ever since.

Retold from Aesop's Fables

1. What did the frog do for the princess?

2. What did the princess promise the frog?

3. Should the king have made the princess keep her promise? Why?

The Frog Prince has been retold many times, both in books and in film. Borrow two different versions of the story from the library and have your child compare and contrast them.

What Time Is It?

Draw lines to match the clocks.

4:00

7:00

5:30

8:30

2:00

3:30

9:00

1:30

Help your child develop a sense of how much time different activities take. Ask questions like, "About how long does it take to brush your teeth—3 minutes or an hour? How long is your favorite TV show—about 30 minutes or 3 hours? How long does it take to eat dinner—30 seconds or 30 minutes?"

We See E's

There are different letter combinations that make the **long e** sound.

e	ee	ea
be	bee	flea
he	see	pea
me	three	sea
we	tree	tea

Use the words in the box to complete the sentences.
Write the words in the puzzle.

Down

1. I drank a cup of ____________ .

2. There is a bird in the ____________ .

3. I can ____________ a bug.

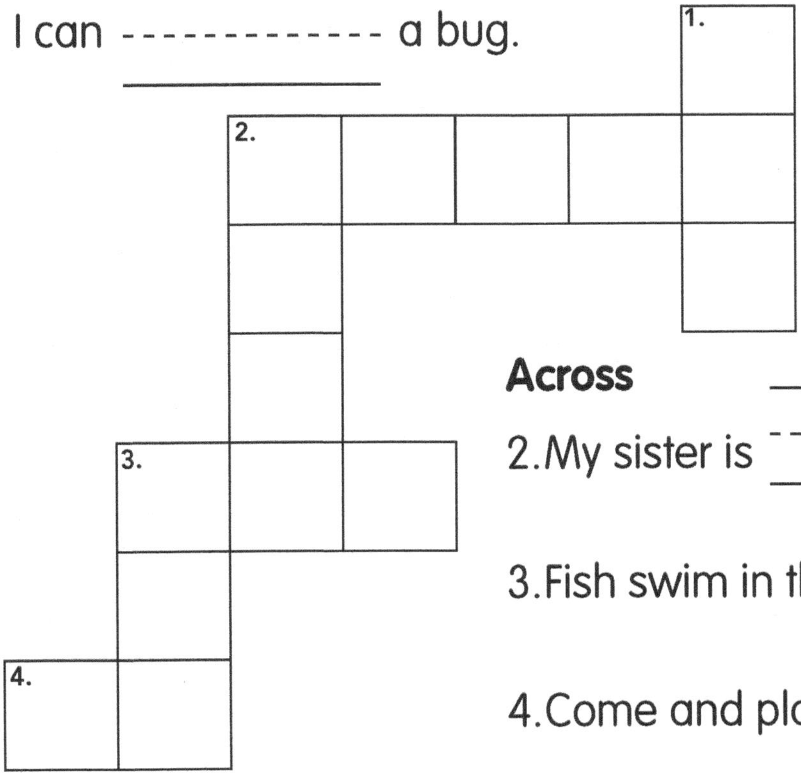

Across

2. My sister is ____________ years old.

3. Fish swim in the ____________ .

4. Come and play with ____________ .

Have your child write sentences for the *long e* words.

Just Like Clockwork

Circle the time shown on each clock.

1.

12:00 3:00 5:00

2.

9:30 8:30 7:30

3.

10:30 11:30 11:30

4.

9:00 6:00 10:00

5.

9:00 3:00 10:00

6.

4:00 2:00 10:00

7.

6:30 12:30 7:30

8.

1:30 3:30 2:30

Have your child show you different times on a toy clock. For each time ask a question such as, "Your clock shows 1:00. What time will it be in two hours?"

Listen to the story. Then answer the questions.

Johnny Appleseed

Jonathan Chapman was born in Massachusetts in September 1774. As a young man, John traveled around the country planting apple orchards. The trees he planted helped the pioneers who moved to the frontier. Johnny was so well-known for his love of apples that people called him "Johnny Appleseed."

Johnny Appleseed had plenty of money, but he lived a very simple life. He slept outdoors and wore scruffy clothes. He walked barefoot, even in snow and ice. Some stories say that he wore a tin pot for a hat. He was very different from most people, but he was also very kind. He helped people and animals everywhere he went.

Today, there are festivals held all around the country to honor Johnny Appleseed and his work.

1. Who is this story about?

2. What was Jonathan Chapman's nickname?

3. What did he do to help pioneers?

Make homemade applesauce with your child in honor of Johnny Appleseed. Peel, core, and chop 3 or 4 apples. Mix the apples with water, sugar, and a dash of cinnamon and cook in a saucepan for 15 minutes. Mash the mixture with a fork. Then cool and serve.

Spare Change

Count the money. Write the amount.

1.

2.

3.

4.

5.

6.

7.

8.

Give your child a handful of coins and ask him or her to show you different ways to arrive at a given amount. For example, 25 cents can be shown with 1 quarter, 2 dimes and 1 nickel, 5 nickels, and so on.

I Spy

There are different letter combinations that make the **long i** sound.

final-e	-y	-ind
hide	by	find
nine	my	kind
time	cry	mind

Read the words in the box. Circle them in the puzzle. They go across and down.

h	i	d	e	a	m	y
e	b	t	i	m	e	f
j	y	n	k	i	f	d
c	g	i	i	n	i	h
r	k	n	n	d	n	e
y	c	e	d	b	d	i

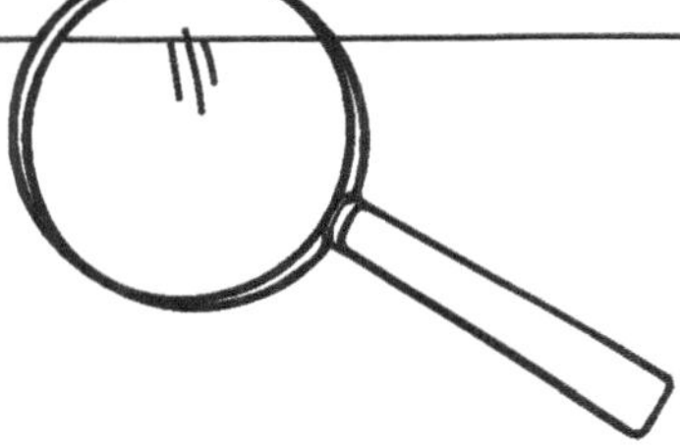

Point out how in some of these words, the *long i* sound is made with *y*. Explain that *y* can also make the *long e* sound, as in *baby*. Make a list of words that end in *y*. Have your child circle the words where the *y* makes the *long i* sound and underline the words with the *long e* sound.

Book Fair

Read the graph. Answer the questions.

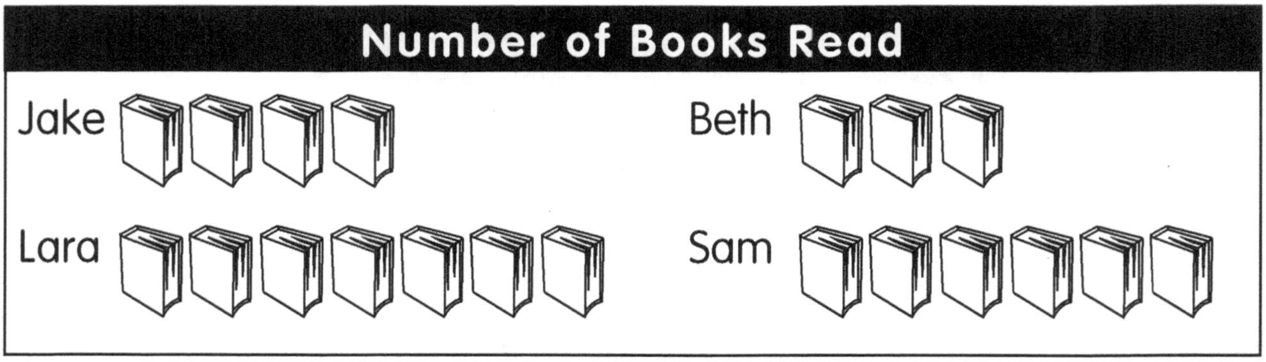

1. Who read the most books? ____________________

2. How many more books did Sam read than Beth? ____________

3. How many books did Jake and Lara read all together? ________

4. Who read the fewest books? ____________________

5. Each student read 10 more books. Fill in the chart to show how many books they read in all.

Jake	
Lara	
Beth	
Sam	

Help your child develop a list of survey questions to ask friends and family and then have him or her graph the results for each question.

Read the poem. Then answer the questions.

Beaver's Pond

Beaver builds a little pond.
He thinks it's just for him.
He fills his home with sticks and mud,
Then takes a little swim.
Along come the spotted deer
To take a nice, cool sip.
Ducks fly high above
Then come in for a dip.
Turtles slide from slippery rocks
And swim among the fish.
Beaver's pond has taken care
Of every animal's wish.

1. Where does this poem take place?

2. What does the beaver use to build his home?

3. Which animals also use the pond?

4. How does Beaver's pond take care of other animals?

Help your child make a shoebox diorama to show life in a pond.

Get Moving!

Complete the chart to find an answer for each problem.

1. Lily ran for 5 minutes on Monday, 10 minutes on Tuesday, and 15 minutes on Wednesday. If she continues this pattern, how long will she run on Saturday? ________

Monday	Tuesday	Wednesday	Thursday	Friday	Saturday
5	10	15			

2. Seth plays basketball. He shot 1 basket in his first game, 4 baskets in his second game, and 7 baskets in his third game. If he continues this pattern, how many baskets will Seth shoot in the sixth game? ________

Game 1	Game 2	Game 3	Game 4	Game 5	Game 6
1	4	7			

3. Lin rode her bike 3 blocks on Thursday, 5 blocks on Friday, and 7 blocks on Saturday. If she continues this pattern, how many blocks will she ride on Tuesday? ________

Thursday	Friday	Saturday	Sunday	Monday	Tuesday
3	5	7			

4. Tomas swam for 10 minutes on Wednesday, 20 minutes on Thursday, and 30 minutes on Friday. If he continues this pattern, how long will he swim on Monday? ________

Wednesday	Thursday	Friday	Saturday	Sunday	Monday
10	20	30			

Write a series of number patterns on a sheet of paper and ask your child to figure out what comes next. Try patterns such as skip counting by twos, threes, fives, and tens.

Listen to the story. Then answer the questions.

Proud Coyote — An Indian Legend

Long ago, an ugly bird jumped into a blue lake. It did this every day for four days. On the fourth day, the bird came out of the lake. Its feathers weren't ugly any more. They were a pretty blue color!

Coyote had been watching all along. He asked the bird, "How did you get such fine feathers?"

The bird said, "Just jump in the lake every day for four days. Then you will have pretty blue fur."

Coyote did just that! On the fourth day he came out of the lake. His fur had turned blue! Coyote was proud of his new color. He wanted everyone to see his fine, blue fur. He walked along, looking left and right to see if anyone was admiring him. Of course, he did not look to see where he was going. He smacked right into an old stump in the ground. He rolled over and over and over in the dust. When he stopped, his fur was no longer blue. It was the color of the dirt. And that is why Coyote is now and forever will be the color of dust and dirt.

1. How did the bird get blue feathers?

2. Why do you think Coyote wanted blue fur?

3. What lesson do you think Coyote learned?

Read this story aloud to your child. Then have him or her read it to you. Choose five words that are difficult for your child to read and write them on index cards. Have your child practice reading those words for a few days. Then come back to this story and see if it is easier for your child to read aloud.

Find the Group

Color the shapes that follow the rule for each group.

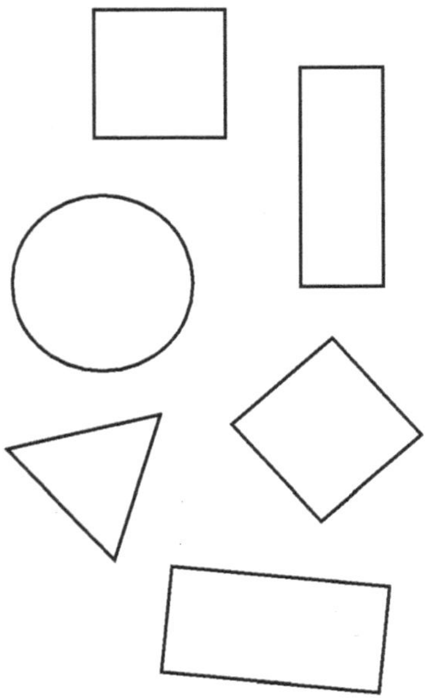

Rule: Shapes with 4 sides

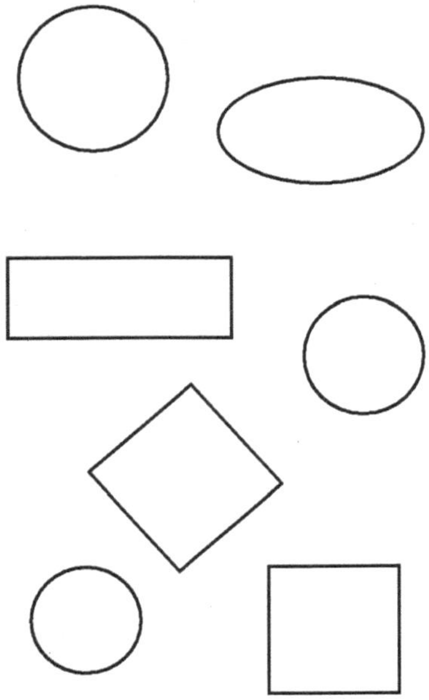

Rule: Shapes with 0 corners

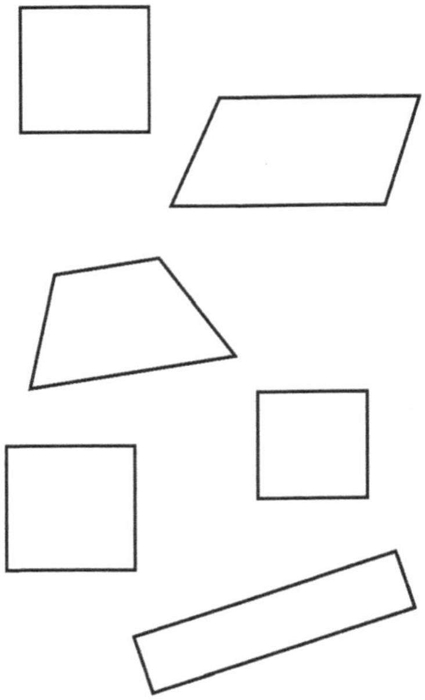

Rule: Squares

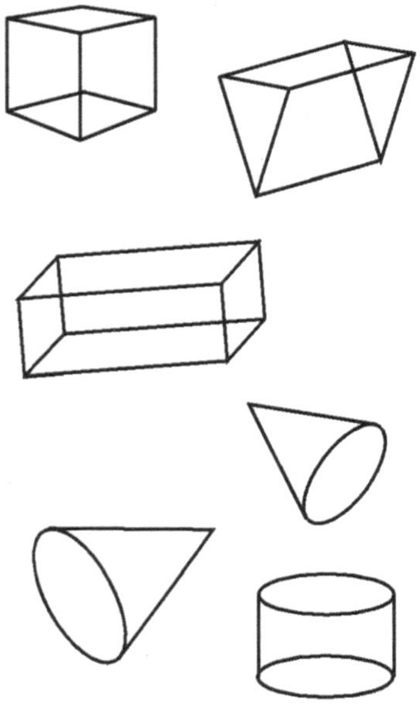

Rule: Cones

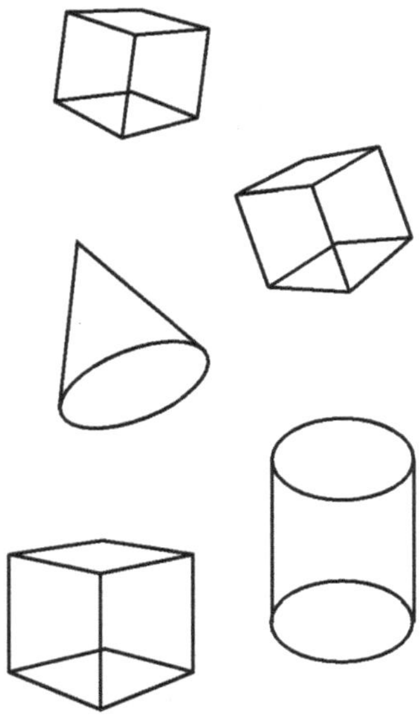

Rule: Cubes

Rule: Shapes with less than 4 corners

Have your child go on a scavenger hunt for shapes around the house. Give clues such as "You can kick and toss this bouncy sphere and find it with your soccer gear." (a soccer ball).

Yo-Ho — It's a Pirate's Life

There are different letter combinations that make the **long o** sound.

o	final -e	ow
go no so	bone joke rope	blow low snow

Circle the misspelled word in each sentence. Write it correctly on the line.

1. Cory told a funny jowk. ______________

2. It was about a pirate in the sno. ______________

3. I'm late, sow I will tell you later. ________

4. How did it goe? __________

5. Oh, noe! __________

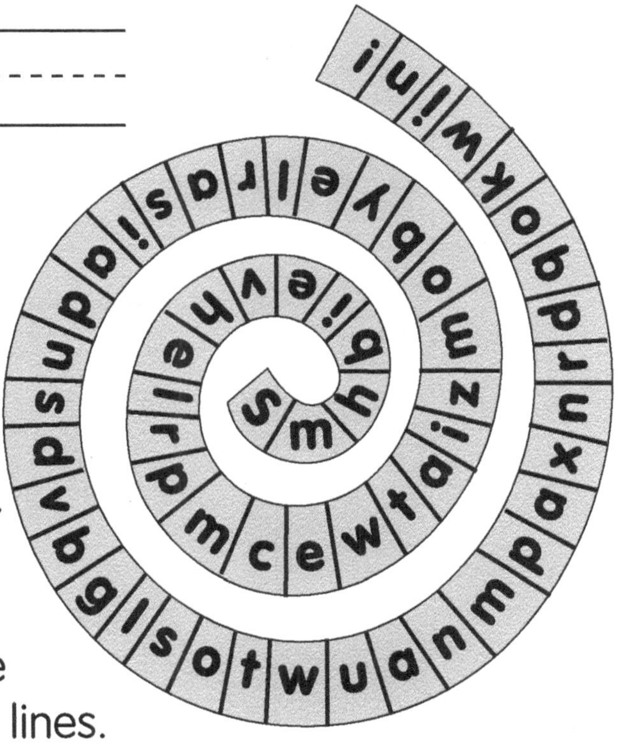

Crack the pirate's code to solve the riddle. Start in the middle of the path. Circle the S. Then skip a letter and circle the next one. Continue this pattern until you reach the end of the path. Write the letters in order on the lines.

What did the pirate say to the snowman on the windy day?

____

Check out a children's riddle book from the library. Have your child read the riddle questions to you and see if you can guess the answers.

Make the Rule

Look at each set of shapes. Sort the set into two groups. Color one group red and the other group blue. Write the rule for each group.

1. 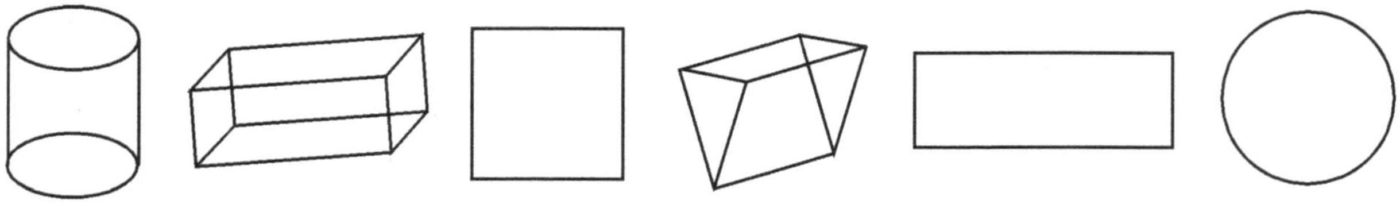

Red Group: ________________ Blue Group: ________________

2. 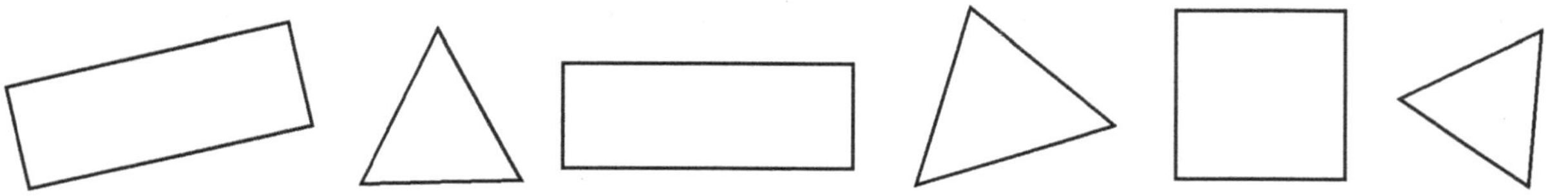

Red Group: ________________ Blue Group: ________________

3. 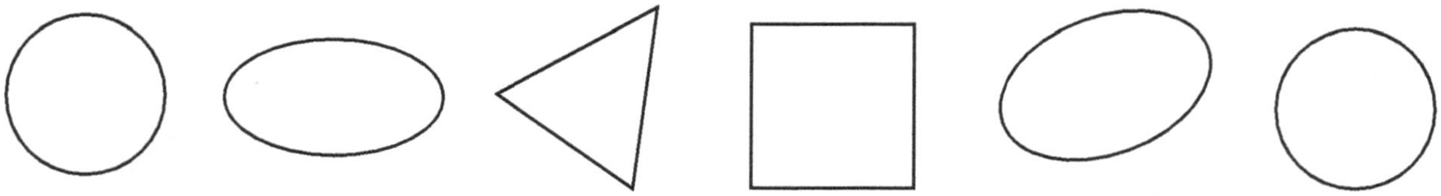

Red Group: ________________ Blue Group: ________________

Have your child select several picture books. Challenge him or her to sort the books into two groups and define what the rule for each group is. Some possibilities include fiction or non-fiction, books about people or animals, and books by certain authors.

Read the passage. Then answer the questions.

Earth's Features

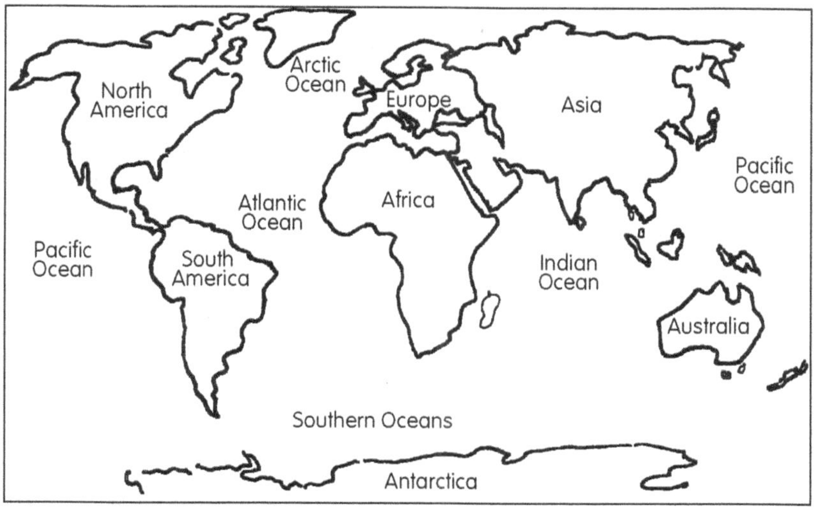

Earth has one giant ocean that covers most of the planet. The ocean is divided into five areas. The areas are named the Atlantic, Pacific, Indian, Arctic, and Southern Oceans.

There are also seven main areas of land on Earth. These are called the continents. People live on only six continents. They are North America, South America, Europe, Africa, Asia, and Australia. Scientists visit the seventh continent, Antarctica, but it is too cold for people to live there all the time.

1. How many oceans does Earth have?

2. What is a continent?

3. Why don't people live on Antartica?

4. Color all the parts of the ocean shown on the map blue.

5. Find your continent on the map. Color it red. Use a different color for each of the other continents.

Have your child use a children's atlas to learn more about the ocean and continents.

Find a Match

Color the shapes that show two matching parts.
Cross out the shapes that do not have matching parts.

1.

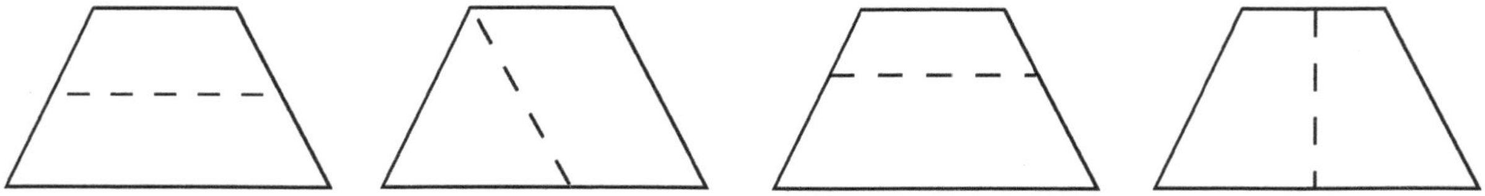

2.

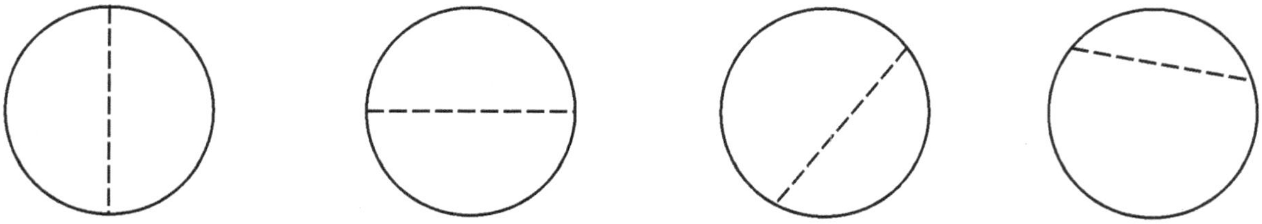

3.

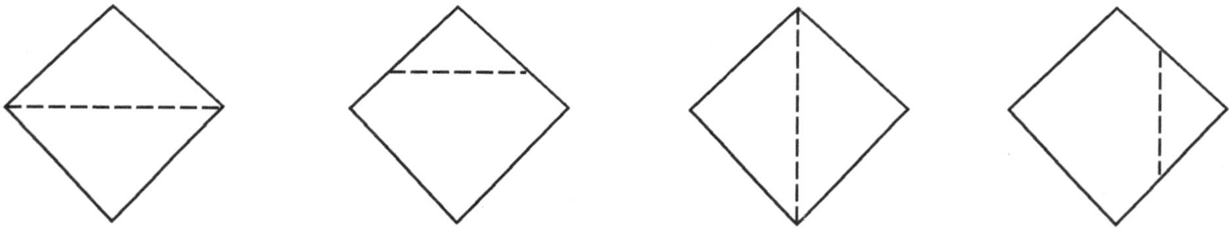

4.

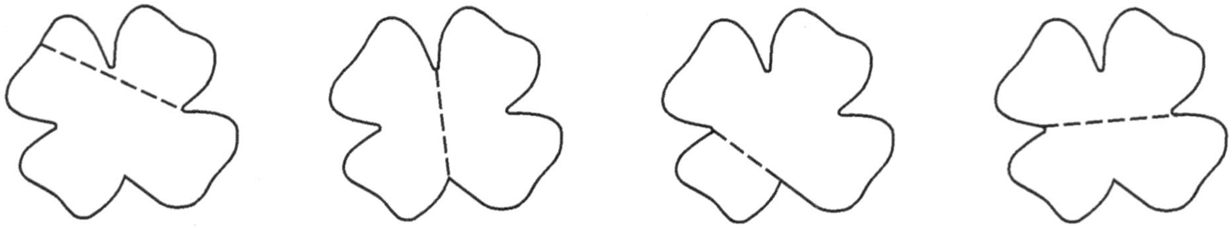

Give your child some soft food that is easily cut, such as tortillas, bread, and sliced cheese. Then have your child use a child-safe knife to cut the food in ways that show symmetry.

Slithery Snakes

Write a word that begins with the letters shown in each section of the snake.

Write two sentences that use words that you wrote on the snakes.

Tongue twisters usually involve a series of repeating sounds. Work with your child to write tongue twisters using the consonant blends on this page. For example, "Stinky Stanley stood on Stubborn Stubby's steps."

Share with Me

Write how many equal shares.

1.

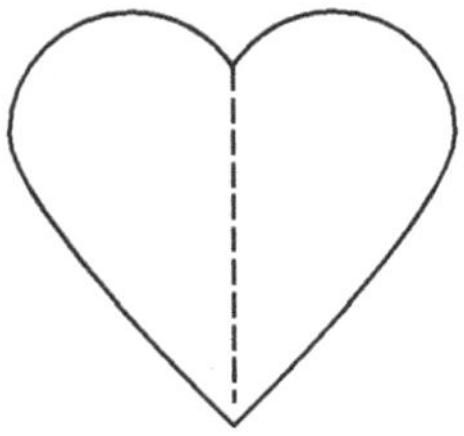

__________ equal shares

2.

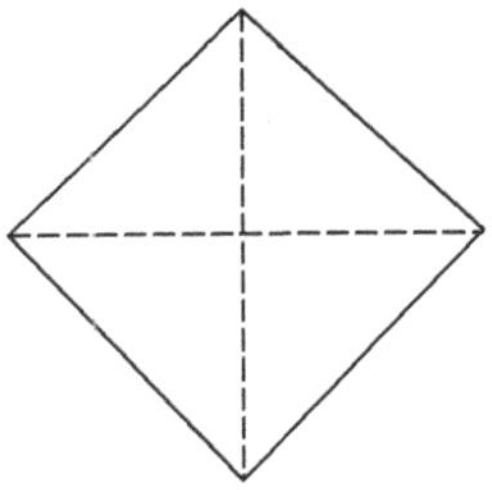

__________ equal shares

3.

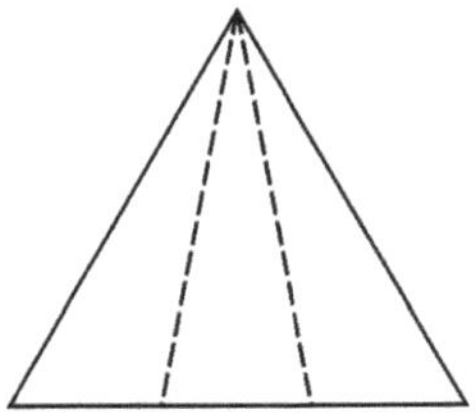

__________ equal shares

4.

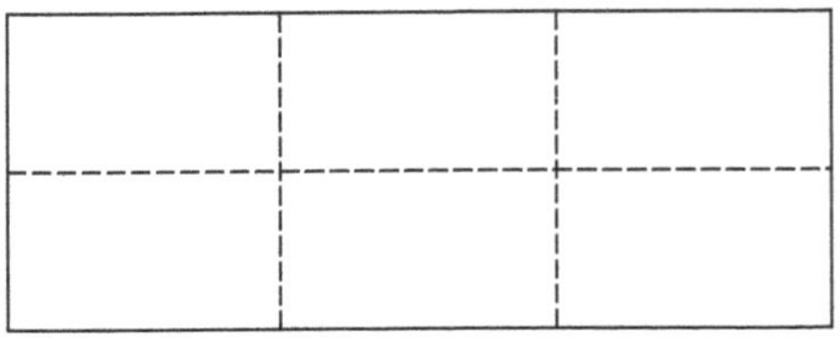

__________ equal shares

5.

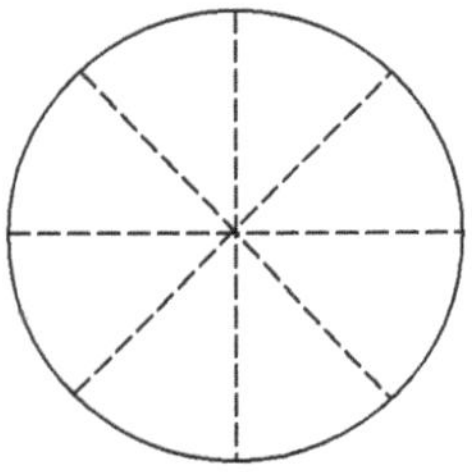

__________ equal shares

6.

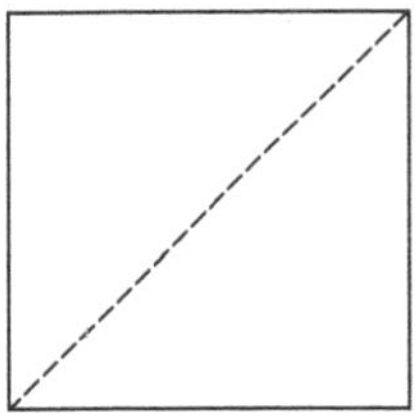

__________ equal shares

Divide the shapes to make more equal shares.

7.

__________ equal shares

8.

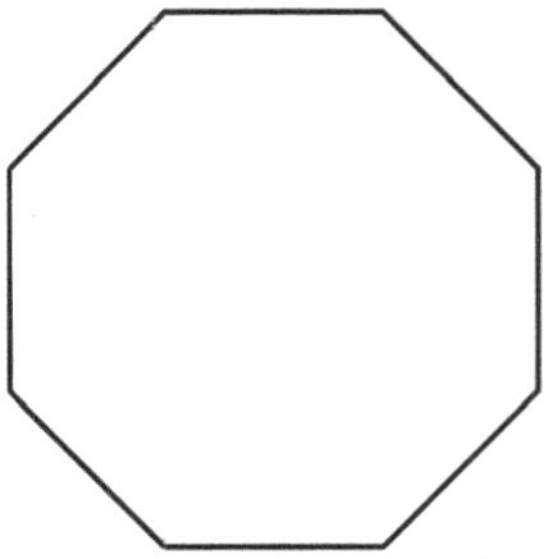

__________ equal shares

Cut various shapes from construction paper and let your child use a marker to divide the shapes into equal shares.

Read the story. Then answer the questions.

Belling the Cat

Long ago, a family of mice lived in a grand house. Everything they wanted was right there in that house. There was fresh bread and rich cheese to eat. There was hot tea and cold milk to drink. There was a cozy fireplace to rest by. The only problem was the cat. He kept the mice trapped in their little hole in the wall. The poor mice could not enjoy anything in the house.

One day, the mice held a meeting. They talked and talked about what to do. At last, a little mouse said, "That cat creeps around as quiet as can be. If we tie a bell around his neck, we will hear him coming. Then we can go where we want and run home as soon as we hear him." The other mice clapped and cheered.

"This is a fine idea," said the old grandmother mouse. "Which of you will tie the bell onto the cat?" The mice looked at each other, but no one said a word. "That's what I thought," said the grandmother. "It is easy to come up with an idea that will never work."

Retold from Aesop's Fables

1. Where does this story take place?

2. Why were the mice unable to have what they wanted?

3. What was the little mouse's idea?

Brainstorm with your child a list of possible solutions to the mice family's problem. Have your child choose one and write a story about it.

Divide It Up

1. Color ½ of each shape.

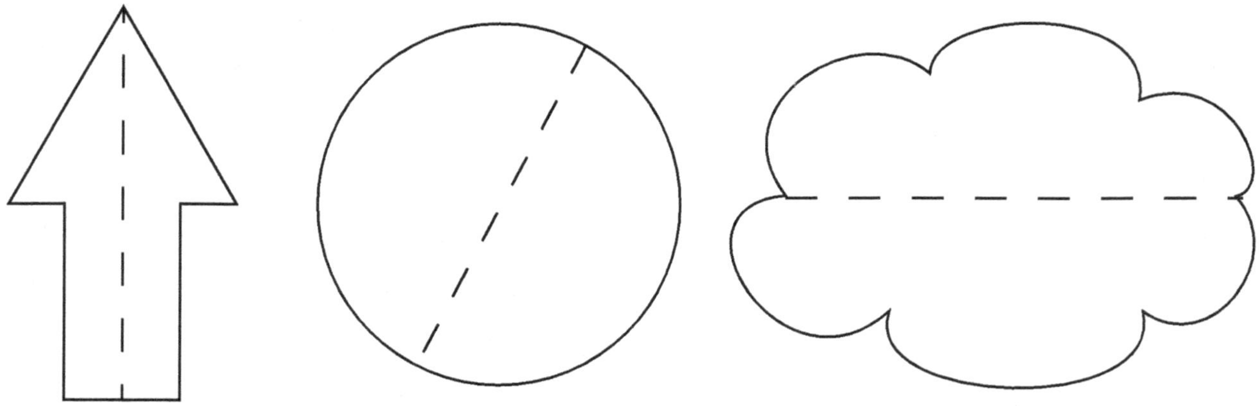

2. Color ¼ of each shape.

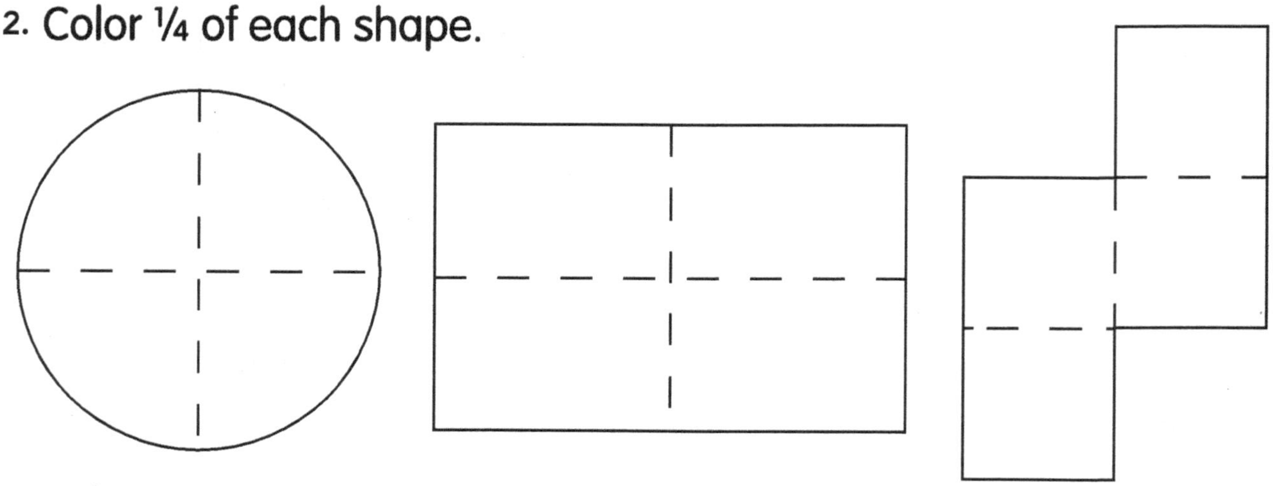

3. Color one part. Circle the fraction.

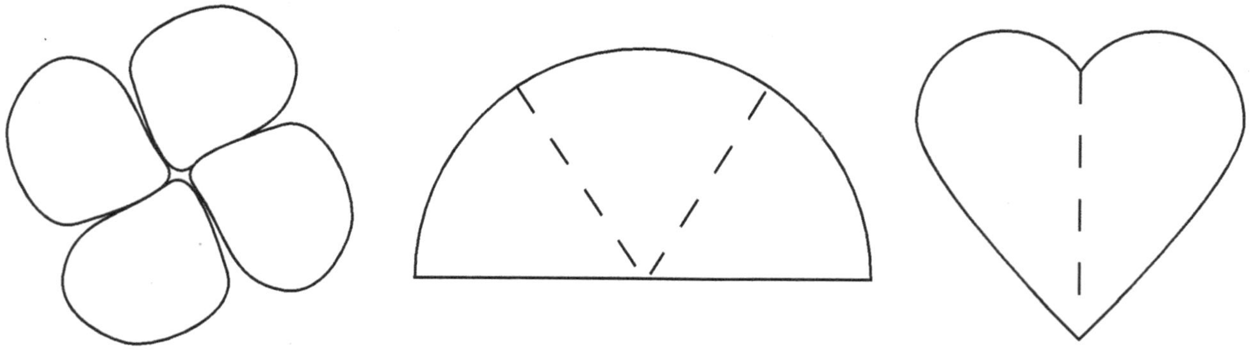

1/2 1/3 1/4 1/2 1/3 1/4 1/2 1/3 1/4

Ask your child how different amounts of food can be divided up equally between groups of friends. For example, if you have 6 slices of pizza and 3 friends, how many slices would each friend have? Encourage your child to draw pictures to show the fractions.

Answer Key

Page 9
1. Little Betty Blue and Little Bo-Peep
2. Answers will vary but may include: Alike—similar names, both lost something; Different—Betty lost a shoe and Bo-Peep lost her sheep, Betty is given a new shoe and Bo-Peep is told to leave her sheep alone.
3. Betty lost a shoe. Bo-Peep lost her sheep.
4. Betty gets a new shoe. Bo-Peep waits for her sheep to return.
5. Answers will vary.

Page 10
1. 3
2. 5
3. 0
4. 1
5. 4
6. 2

Page 12
1. 10
2. 7
3. 9
4. 6
5. 8

Page 13
1. 13
2. Because there are 50 states
3. Red, white, and blue

Page 14
1. 12
2. 15
3. 11
4. 14
5. 13

Page 16
1. 18
2. 20
3. 16
4. 19
5. 17

Page 17
1. Rum, pum
2. Colin
3. The band wants to play and have us dance all day.

Page 18
1. 10, 11, 12, 13
2. 4, 5, 6, 7
3. 16, 17, 18, 19
4. 12, 13, 14, 15

Page 19
1. A picnic
2. Milk
3. Answers will vary.

Page 20
1. 10, 20, 30, 40, 50, 60, 70
2. 10, 20, 30, 50
3. 10, 30, 40, 50, 60 70, 80, 90
4. 10, 20, 40, 50, 60, 70, 80, 90, 100
5. 20, 30, 40
6. 10, 20, 30, 40, 60, 70, 80

Page 21
Answers will vary. May include these words: cat—bat, hat, fat, etc; fan—ran, Dan, man, etc; had—sad, mad, dad, etc; big—dig, fig, pig, etc; sit—bit, hit, lit, etc; lip—clip, dip, flip, etc.

Page 22
1. 5, 10, 15, 20, 25
2. 5, 10, 15
3. 5, 10, 15, 20, 25, 30, 35, 40
4. 5, 10, 15, 20, 25, 30, 35
5. 5, 10, 15, 20, 25, 30

Page 23
1. Their own food
2. Their leaves
3. They use their roots to take in water.
4. Clothes

Page 24
1. 2, 4, 6, 8
2. 2, 4, 6, 8, 10, 12, 14, 16
3. 2, 4, 6, 8, 10
4. 2, 4, 6, 8, 10, 12, 14, 16, 18, 20, 22

Page 25
Answers will vary but may include the following:
1. dog, jog, log
2. Ben, den, hen
3. bun, fun, sun
4. got, hot, lot
5. bug, dug, rug

Page 26

1	2	3	4	5	6	7	8	9	10
11	12	13	14	15	16	17	18	19	20
21	22	23	24	25	26	27	28	29	30
31	32	33	34	35	36	37	38	39	40
41	42	43	44	45	46	47	48	49	50
51	52	53	54	55	56	57	58	59	60
61	62	63	64	65	66	67	68	69	70
71	72	73	74	75	76	77	78	79	80
81	82	83	84	85	86	87	88	89	90
91	92	93	94	95	96	97	98	99	100

Page 27
1. Logan
2. at his home. Words in the story that should be underlined include: chair, table, closet, bed.
3. his toy dog
4. under the bed

Page 28
1. 17
2. 34
3. 23, 25
4. 71, 73
5. 83, 85
6. 15, 17, 18
7. 61, 64, 66
8. 39, 40, 43

Page 29
1. Jill and hill, down and crown
2. to get water
3. Answers will vary. It means he broke his head.

Page 30
1. 3
2. 7
3. 9
4. 8
5. 10
6. 5
7. 2
8. 4

Page 31
1. g
2. l
3. t
4. b
5. r
6. n
7. p
8. m

Page 32
1. 6
2. 2
3. 2
4. 4
5. 3
6. 3

Page 33

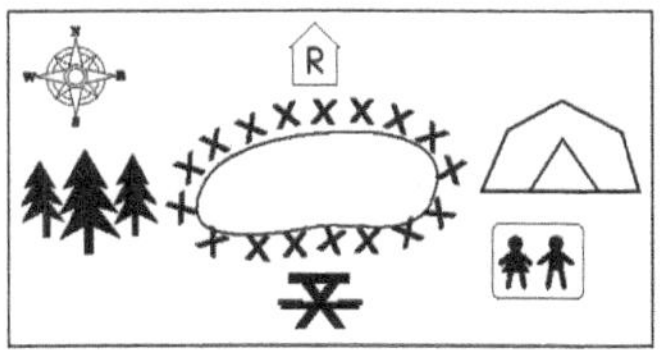

Page 34
Child should draw enough apples on the tree to make 10 apples in all. Addition problems should be written as follows:
1. 5 + 5 =10
2. 7 + 3 = 10
3. 1 + 9 = 10
4. 4 + 6 = 10
5. 8 + 2 = 10
6. 0 + 10 = 10

Page 35
1. e
2. u
3. i
4. o
5. a
6. u
7. i
8. a

Page 36
1. 6
2. 4
3. 5
4. 4
5. 4
6. 1

Page 37
1. soccer
2. 9:30
3. Ben's sister likes to sing.

Page 38
1. 5
2. 7
3. 6
4. 10

Page 39
1. at the park
2. She went down the slide.
3. She ate lunch.

Page 40

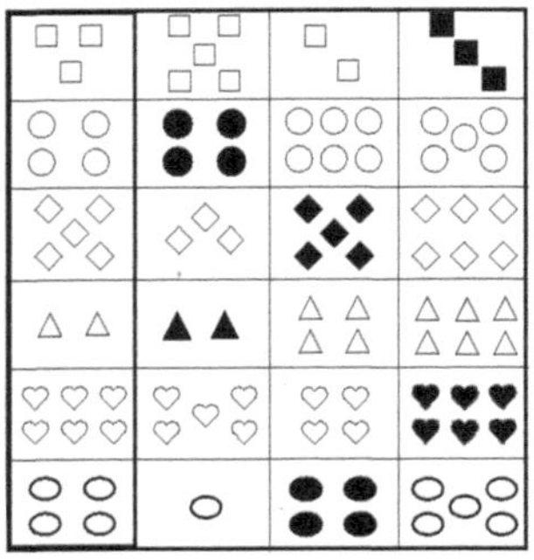

Page 41
1. to
2. down
3. in
4. will, she
5. go, for

Page 42
1. 11, 15
2. 16, 13
3. 20, 17
4. 10, 12
5. 18, 14
6. 19, 16

Page 43
1. a pool of water with land all around it
2. turtles, fish, and frogs
3. to drink water and to look for food

Page 44
Child should draw a set of pictures smaller than the set shown.
1. 6
2. 8
3. 2
4. 9
5. 7
6. 4

Page 45
1. and
2. do
3. are
4. want
5. see
6. see
7. are
8. want
9. do
10. and

Page 46

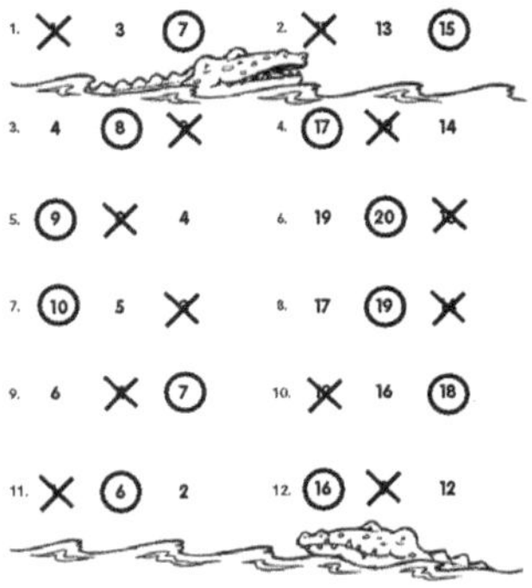

Page 47
1. to the store
2. cake mix, candles, and frosting
3. a birthday
4. because May and her dad will bake her a cake.

Page 48

Page 49
1. at the library
2. It was funny.
3. The puppets danced and told jokes.
4. make her own puppets

Page 50
1. 1 dime, 2 pennies; 12 pennies
2. 1 dime, 0 pennies; 10 pennies
3. 1 dime, 5 pennies; 15 pennies
4. 1 dime, 7 pennies; 17 pennies
5. 1 dime, 4 pennies; 14 pennies
6. 1 dime, 6 pennies; 16 pennies

Page 51
Answers will vary.

Page 52
1. 10 + 3 = 13
2. 10 + 5 =15
3. 10 + 9 = 19
4. 10 + 2 = 12
5. 10 + 4 = 14
6. 10 + 7 = 17

Page 53
1. 2007
2. She rode a pony.
3. 3, 1, 2

Page 54
The first word below shows the picture that should be crossed out. The next word defines the remaining group of pictures:
1. fish, insects (or bugs)
2. monkey, farm animals
3. cereal, fruit
4. tree, things in the sky
5. skateboard, vehicles
6. globe, art tools

Page 55

Page 56

Page 57
1. a turnip
2. It grew too big.
3. His family and pets
4. teamwork

Page 58
1. 8:00
2. 12:00
3. 3:00
4. 6:00
5. 9:00

Page 59
1. to the movies
2. The lights went out.
3. They loved it.

Page 60
1. 7, 7, 12, 12
2. 13, 8, 7, 13
3. 20, 10, 12, 14
4. 20, 17, 17, 19
5. 19, 20, 19, 20

Page 61
Answers will vary.

Page 62
Answers will vary, but may include the following:
1. 10 + 7, 16 + 1, 9 + 8
2. 10 + 2, 11 + 1, 6 + 6
3. 10 + 5, 14 + 1, 12 + 3
4. 10 + 4, 13 + 1, 7 + 7
5. 5 + 5, 6 + 4, 7 + 3
6. 10 + 6, 15 + 1, 8 + 8
7. 10 + 9, 18 + 1, 15 + 4
8. 10 + 1, 5 + 6, 7 + 4

Page 63
1. reptiles
2. Possible answers include snakes, lizards, turtles, and alligators
3. to warm up
4. Reptiles are interesting creatures.

Page 64
1. 12
2. 16
3. 19
4. 11
5. 13
6. 15
7. 20
8. 17
9. 19

Page 65
1. Rabbits hop.
2. I swim.
3. We sing.
4. He bakes.
5. She runs.
6. Cats nap.
7. They sit.
8. You play.
Sentences will vary.

Page 66
a. 3
b. 7
c. 4
d. 10
e. 11
i. 6
l. 2
o. 5
s. 12
t. 1
u. 8
w. 9
because it was too colt outside!

Page 67
1. Pretend, because there is magic in it.
2. Answers will vary.
3. Answers will vary.

Page 68
1. 11
2. 13
3. 7
4. 17
5. 13
6. 20

Page 69
1. excited
2. shy or timid
3. Answers will vary.

Page 70
1. 7, 9, 3
2. 6, 0, 8
3. 10, 10, 5
4. 3, 10, 5
5. 10, 9, 5
6. 12, 8, 4, 11, 3
7. 15, 12, 4, 10, 5

Page 71

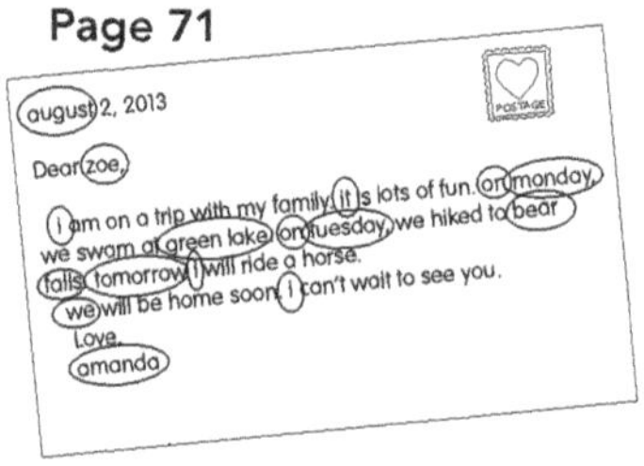
august 2, 2013
Dear zoe,
I am on a trip with my family. it is lots of fun. on monday, we hiked to bear falls. tomorrow I will ride a horse.
we swam at green lake on tuesday.
we will be home soon. I can't wait to see you.
Love,
amanda

Page 72

Page 73
1. Africa
2. 2
3. Rivers
4. 17
5. Glossary

Page 74
1. 2, 2, 2
2. 3, 3, 3
3. 2, 2, 2
4. 1, 1, 1
5. 5, 5, 5
6. 4, 4, 4
7. 6, 6, 6
8. 7, 7, 7

Page 75
1. .
2. ?
3. !
4. .
5. !
6. !
7. .
8. ?
Sentences will vary.

Page 76
1. 6
2. 14
3. 16
4. 4
5. 8
6. 10
7. 12
8. 4

Page 77
1. a man
2. so he could have his bone all to himself
3. his own reflection

Page 78
1. 5
2. 4
3. 6
4. 10
5. 4
6. 7

Page 79
1. Brenna
2. hopping from flower to flower
3. wings and a dress
4. a flower fairy
5. Answers will vary.

Page 80
1. 87, 88, 89, 90, 91, 92
2. 55, 56, 57, 58, 59, 60
3. 99, 100, 101, 102, 103, 104
4. 65, 70, 75, 80, 85, 90
5. 80, 85, 90, 95, 100, 105
6. 90, 95, 100, 105, 110, 115
7. 60, 70, 80, 90, 100, 110
8. 20, 30, 40, 50, 60, 70
9. 50, 60, 70, 80, 90, 100

Page 81
1. can (cane)
2. (kite) kit
3. (hope) hop
4. cub (cube)

Page 82
1. 100, 92, 119
2. 86, 110, 69
3. 109, 103, 119
4. 116, 79, 55
5. 99, 108
6. 112, 114, 118
7. 90, 105, 110

Page 83
1. true
2. false
3. true
Pictures will vary.

Page 84
1. 32
2. 20
3. 59
4. 1 ten, 7 ones
5. 4 tens, 0 ones
6. 3 tens, 5 ones

Page 85
1. mail
2. stay
3. late
4. gray
5. pail
6. rain

Page 86
1. >, <, <
2. =, >, >
3. <, <, >
4. <, >, >
5. >, =, <

Page 87
1. long ago
2. It was the first day of school.
3. She knew all her letters.
4. Answers will vary.

Page 88
1. 33, 24, 49
2. 50, 86, 91
3. 42, 58, 85
4. 16, 31, 27

Page 89
1. got her ball for her
2. to be his friend
3. Answers will vary.

Page 90

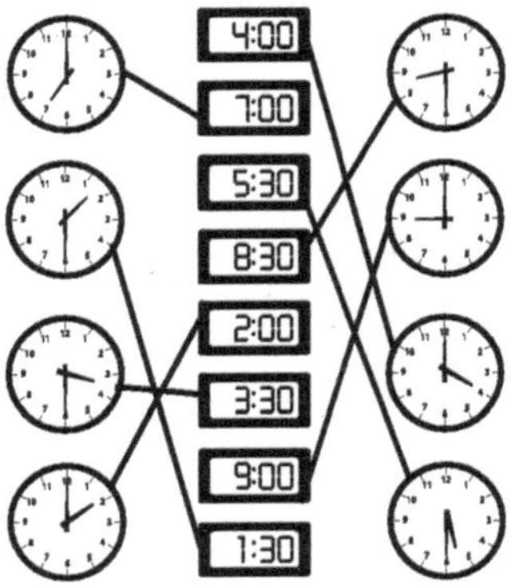

Page 91

Down
1. tea
2. tree
3. see

Across
2. three
3. sea
4. me

Page 92
1. 12:00
2. 7:30
3. 11:30
4. 6:00
5. 9:00
6. 10:00
7. 6:30
8. 2:30

Page 93
1. Jonathan Chapman or Johnny Appleseed
2. Johnny Appleseed
3. planted apple trees

Page 94
1. 8¢
2. 14¢
3. 20¢
4. 25¢
5. 6¢
6. 15¢
7. 16¢
8. 17¢

Page 95

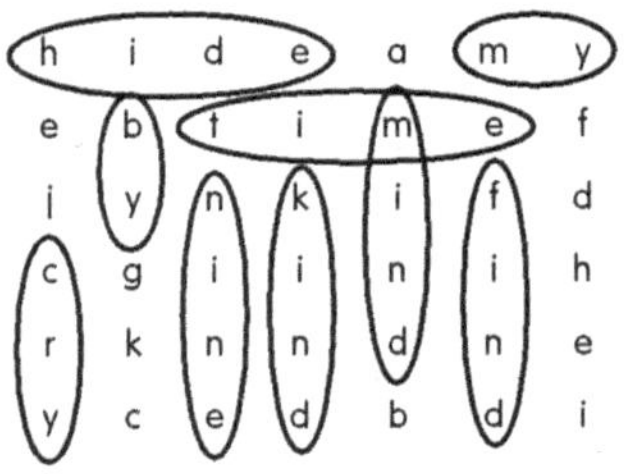

Page 96
1. Lara
2. 3
3. 11
4. Beth

Jake	14
Lara	17
Beth	13
Sam	16

Page 97
1. at a pond
2. sticks and mud
3. deer, ducks, turtles, fish
4. It gives them water.

Page 98
1. 30 minutes
2. 16 baskets
3. 13 blocks
4. 60 minutes

Page 99
1. It jumped in the lake.
2. He thought it would be pretty.
3. Answers will vary.

Page 100

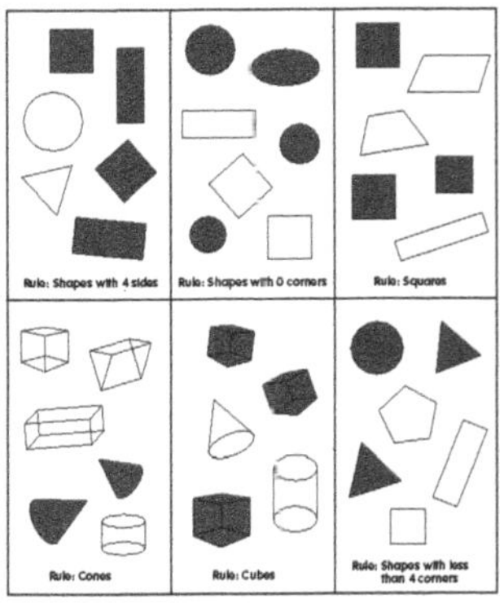

Page 101
1. jowk—joke
2. sno—snow
3. goe—go
4. sow—so
5. noe—no

Shiver me timbers and blow a man down!

Page 102
Answers will vary.

Page 103
1. one
2. a main area of land on Earth
3. It is too cold.

Page 104

Page 105
Answers will vary, but may include the following:
1. slam, sled, slide, slip
2. smile, smell, smoke, smack
3. snack, snail, snip, snow
4. stop, step, stay, star

Page 106
1. 2
2. 4
3. 3
4. 6
5. 8
6. 2

Answers will vary for problems 7 and 8.

Page 107
1. in a house
2. The cat kept the mice in their hole.
3. to put a bell on the cat

Page 108

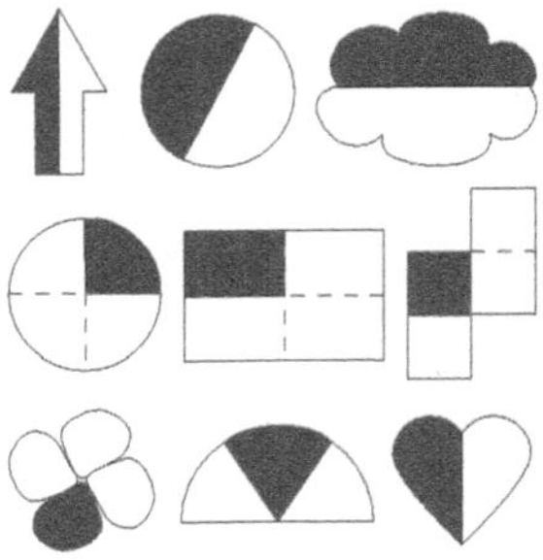

$\frac{1}{4}$ $\frac{1}{3}$ $\frac{1}{2}$